CITY STEPS
of PITTSBURGH

CITY STEPS
of PITTSBURGH
A HISTORY & GUIDE

LAURA ZUROWSKI, CHARLES SUCCOP AND MATTHEW JACOB
Foreword by Bob Regan

THE
History
PRESS

Published by The History Press
Charleston, SC
www.historypress.com

First published 2024

Manufactured in the United States

ISBN 9781467156721

Library of Congress Control Number: 2024931455

We walk and document these lands, but they are not ours.

The authors of this book wish to honor the land and the resilient creatures who are our neighbors here. We acknowledge this region as the ancestral home of the Adena Culture, Hopewell Culture and Monongahela peoples, as well as the Delaware, Shawnee and Haudenosaunee peoples who also lived here. In many places, our city steps have followed the time-worn traveling paths created by these original inhabitants. They were the first to experience the challenges of this unruly terrain and devise efficient and effective means of traveling throughout. We hope that their commitment and contributions to living in harmony with the natural world will inspire others to do the same.

CONTENTS

CONTENTS

FOREWORD

For nearly a century, Pittsburgh created municipal infrastructure to transport a consistently growing population that ultimately housed 700,000 people. However, within a few decades of that apex, the city lost nearly 60 percent of its population and became yet another "Rust Belt" metro struggling to stay afloat in a very different economic climate. This is the Pittsburgh I encountered in the late 1990s, when I relocated from Boston and began teaching at the University of Pittsburgh in the then-emerging field of geographic information systems (GIS). Soon after arriving, I became intrigued by the stairs and wanted to know more, but I was surprised to discover that records were few and far between. So, I set out on my bicycle with the ambitious goal of locating and mapping all of Pittsburgh's city steps. I was soon joined in my quest by Tim Fabian, who photographed the steps. Our mission, which to the best of our knowledge had never been done before, was to catalogue and document the steps. As a result of our efforts, we ultimately located 739 distinct flights in sixty-six of Pittsburgh's ninety neighborhoods and launched efforts to publicize them with the hopes of fostering their preservation. This work culminated in the publication of *The Steps of Pittsburgh: Portrait of a City* in 2004, which quickly sold out and went out of print.

But my connection with the city steps was more than locating and mapping them; I viewed the steps as a unique and wondrous asset that the city had failed to recognize, preserve and promote. Unfortunately, municipal support for the stairs waxed and waned in the years following that book, even though the subject consistently garnered publicity through various news stories. One such story, a front-page article in the *Wall Street Journal* in 2014 by John Miller, resulted in the publication of *Pittsburgh Steps: The Story of the City's Public Stairways* (2015), which gave a new life to the original book. Within a year, the data I had collected for the book was incorporated into the City Steps Plan of Pittsburgh's newly created Department of Mobility and Infrastructure. This plan, which involved substantial community input, resulted in a comprehensive approach to determining which flights were most essential to neighborhood mobility and in need of funding for repairs. While change within a city government rarely comes quickly, it's reasonable to say that the current level of interest in preserving the stairs is nearly as great now as it was in the post–World War II years during Pittsburgh's population expansion.

In recent years, we've seen a strong resurgence of interest in the city steps, with residents and elected officials advocating for additional funding for their upkeep and repair. This new movement is due in no small part to Laura Zurowski, who moved to Pittsburgh in 2011 and soon became intrigued by our city steps, just as I was. She began a documentary project, Mis.Steps (mis-steps.com), to photograph and write about every flight of steps in the city. Gifted with abundant energy, Laura not only revisited the stairs listed in the *Pittsburgh Steps* book but also set her sights on documenting the flights not included in my original work, including more than one hundred flights of stairs in Pittsburgh's public parks. In addition to these contributions, she regularly shares her knowledge through public walking tours and talks about the history of the steps.

Twenty years have passed since my first book on the steps, and much has changed in Pittsburgh. The time is right for new exploration on this important topic. Laura has marshaled a team of steps enthusiasts to produce an exciting, informative and illustrative new book on Pittsburgh's city steps that is a must-read for all Pittsburghers and anyone interested in these unique artifacts of our city.

—BOB REGAN

PREFACE

We didn't know one another in 2017, but Matthew Jacob, Charles Succop and I were traveling distinct paths that ultimately brought us to a point of intersection.

After consulting with author and Pittsburgh steps enthusiast Bob Regan, I had recently started re-creating his urban exploration with an Instagram-based documentary project called Mis.Steps: Our Missed Connections with Pittsburgh Public Stairways. I had recently moved to Pittsburgh from Providence, Rhode Island, and was captivated by the landscape, architecture and industrial history that were, in turn, pleasantly familiar yet strikingly different. Matthew was working for Pittsburgh's Department of Public Works, creating the city's first digital database of municipal infrastructure assets, such as retaining walls, streetlights, ADA curb cuts, bridges and city steps. Matthew was tasked with incorporating Regan's city steps data gathered during his travels throughout Pittsburgh during the late 1990s (with his permission) into that database and photographing current conditions. Charles had recently started working in the newly established Pittsburgh City Archives and was embarking on the monumental task of processing more than two hundred years' worth of municipal records.

Throughout the next five years, Instagram grew in popularity, and Matthew, Charles and I "knew" one another through our online profiles—I was walking the city's neighborhoods for Mis.Steps, Matthew was sharing historical information about the city steps and mapping trails throughout the city's greenspaces and Charles was creating meticulously researched "slice of life" posts through his Pittsburgh Then and Now account.

In 2021, after three years of walking through Pittsburgh's neighborhoods and writing about my experiences traveling the city steps, I realized that the time was ripe for creating a new book—Pittsburgh was a very different place from the one Bob Regan had encountered in the late 1990s. Soon after, Matthew and Charles signed on, and our team was born. Each of us has contributed in unique ways to this project: I have walked the city steps and streets of nearly every neighborhood in Pittsburgh, plotted the walking tours and written the histories and stories told throughout this book. Charles conducted historical research using the City of Pittsburgh Archives and other publicly available sources. Matthew created the tour maps and took the contemporary photos that appear in the book. This division of labor aligned with our strengths and our interests.

While each of us could have created this book independently, we believe it is more substantial and compelling because of our collaboration. To quote John C. Maxwell, "Teamwork makes the dream work," but we must acknowledge that this dream of creating a brand-new Pittsburgh city steps book wouldn't have been possible without the work of Bob Regan more than twenty years ago.

This book represents the Pittsburgh we find today in 2024. Our aim as writers, researchers and documentarians has been to create a book about the history and importance of this very old and original form of infrastructure that is still highly relevant today. The city steps have a role to play in this new Pittsburgh, a Pittsburgh that has a greater understanding of the past and more options than ever before for the future. We hope that this book and the walks it includes inspire readers to get out and re-engage with the city steps and neighborhoods by slowing down, walking and observing. Our highly connected, instantaneous information age provides us with answers to nearly any question, but it can never take the place of in-person, feet-on-the-ground exploration.

The city steps were built to move people within and between neighborhoods. By following these paths, it is possible to walk alongside the past and present simultaneously. That experience is what Pittsburgh needs to create its best possible future.

—LAURA ZUROWSKI

ACKNOWLEDGEMENTS

First and foremost, we would like to express our deepest gratitude to Bob Regan, who has been a source of guidance, inspiration and encouragement throughout the writing of this book. Without his support and passion for the city steps, this book would not have been possible.

We are also grateful to J. Banks Smither and The History Press for enthusiastically embracing our manuscript. Their unwavering dedication, superior communication skills and positive "can do" attitude have been truly appreciated. Working with you has been an absolute joy!

Special thanks are due to our community reviewers Scott Bricker, David Kidd, Sue Morris, Rich Cogden and Lesley Rains. Their passion and knowledge of the city steps and Pittsburgh's history added valuable feedback and interesting details to the text. Thanks also to our "test walker," Marcher Arrant, who ensured that the technical directions for every walk were accurate and easy to navigate. The stories, information and experience of walking Pittsburgh's many neighborhoods are better because of their time and dedication.

Additionally, we'd like to thank Patrick Hassett and Thomas Joyce for their interviews and conversations about the history of Pittsburgh's Department of Public Works. Project Manager Brian Masci and the crew at BCI Commercial Concrete provided all the details about the Copperfield Avenue rebuild project and were always good sports about answering questions. We also wish to acknowledge and thank the many decades of engineers, skilled tradesmen and women, laborers and city employees who have been responsible for constructing and maintaining the city steps. Their names may be unknown, and their work underappreciated, but their skill and dedication are the foundation of this book.

Spending time with Anthony J. Mainiero, "Mr. Tony," was an unforgettable highlight of writing this book. Our lives are so much richer because of your life and stories, Tony. Thank you also to John Peña for facilitating our early conversations.

We especially appreciate the following sources for several of our images: Pittsburgh City Photographer Collection, 1901–2002, AIS.1971.05, Archives & Special Collections, University of Pittsburgh Library System, and the Carnegie Museum of Art, Heinz Family Fund, © Carnegie Museum of Art, Charles "Teenie" Harris Archive.

FURTHERMORE, WE WOULD LIKE to acknowledge the following people and organizations that have contributed to our work of promoting the city steps and bringing this book to fruition: Allegheny Cleanways, Atlas Obscura, BikePGH, Carnegie Museum of Art's Charles "Teenie" Harris Archive, City Cast Pittsburgh, the Thomas and Katherine Detre Library & Archives at the Senator John Heinz History Center, Doors Open Pittsburgh, Garden Club of Allegheny County, *Governing* magazine, Hilltop Coffee, Monroeville Public Library, Osher Lifelong Learning Institute at Carnegie Mellon University, Pittsburgh City Archives Records Management Division, Pittsburgh City Clerk's Office, *Pittsburgh Magazine*, *Pittsburgh Orbit*, Society for Contemporary and Historical Archaeology in Theory (CHAT), Society for Industrial Archaeology, Squirrel Hill Historical Society, Threadbare Cider, Union Fitness, University of Pittsburgh Archives & Special Collections, Velum Fermentation, Venture Outdoors, Walk the Burgh Tours, 90.5 WESA, Chelsea Arthur, Avi, Eric Bean, Jason Armstrong Beck, Dean Bog, Seth Bush, Brian Butko, Alyssa Cammarata Chance, Leslie Clague, Umberto Crenca, Rossilynne Culgan, Ron Donoughe, Catherine Drabkin, Erin Feichtner, Elizabeth Fein, Bill Fink, Miranda Gard, Kevin Gavin, Meredith Grelli, Nils Hanczar, Nick Hartley, Russ Hexter, Miles Howard, Maura Jacob, Robert Jacob, Kelly Klabnik, Ellen Kotzin, Simeon Larivonovoff, Haley LeFrancois, Hope LeVan, Cory Little, Edward May, Miriam Meislik, Alden Merchant, Elise Miranda, Bradford Mumpower, Mary O'Loughlin, John O'Nan, Stewart O'Nan, Bill Peduto, Stuart Putnam, Jimmy Riordin, Diane Samuels, Valentina Scholar, Travis Straub, Beth and Gus Succop, Giovanni Svevo, Victor Van Carpels, Sarah and Gary Weiss and Eric Younkins.

Last but not least, our heartfelt thanks to all the readers, walkers and explorers who have taken the time to engage with the city steps. Your interest and enthusiasm are what make Pittsburgh an amazingly wonderful and welcoming place.

• • • • • •

"The life and passion of a person leaves an imprint on the ether of a place. Love does not remain within the heart, it flows out to build secret tabernacles in a landscape."

—John O'Donohue

HOW TO USE THIS BOOK

Pittsburgh's public stairways are integral to the city's history and are a point of civic and neighborhood pride. Today, their presence provides a uniqueness that sets it apart from other cities and offers visitors and locals a built-in way to travel the city, get exercise and connect with other residents.

City Steps of Pittsburgh: A History & Guide is a guidebook designed for longtime locals, new residents and visitors to learn more about the city steps and explore neighborhoods in all areas of the city. In addition to offering a historical overview of the stairs and their role in the lives of ordinary Pittsburghers, this book also provides eleven different guided city steps walks and individual locations unique because of their length, steepness or "off-the-beaten-path" qualities. The walks include maps that take readers on journeys through the past and present. This book is intentionally light, compact and fun, with interesting information that isn't too weighty (literally and figuratively). Our approach is that of a trusted friend who takes you to see some of the best features of their favorite neighborhoods—it's conversational, nontechnical and informative. However, for those who want to dig deeper into topics touched on in this book, a selection of books, maps and website resources is included at the end.

Each walk explores a different Pittsburgh neighborhood and stairway route. Guided tours include an overview of the neighborhood's history and its role in Pittsburgh's development, as well as a few standard details for quick reference:

LENGTH: Expressed in miles; all walks loop around unless otherwise noted.

LEVEL OF DIFFICULTY: Subjective according to ability. Pittsburgh's terrain is naturally challenging, and the amount of effort required depends on the intensity and duration of elevation changes.

GETTING THERE: How to access routes by vehicle, public transit, on foot or by bike.

STREET BOUNDARIES: Provided for walkers who may wish to enter/exit the route at a different location or conduct personal explorations into nearby areas.

ROUTE DESCRIPTION: Full directions, described in segments. Each segment references a numbered site on the map so you can match directions and places. Occasionally, supplemental routes (side trips from the main route) are provided for further exploration.

MAPS: Follow the site number sequence and directions. The locations of the stairs are shown, including starting and ending points.

INSIDER TIPS FOR THE BEST ADVENTURE

Urban hiking requires precautions, and safety is always our first priority. Only you can keep yourself safe; stay alert when crossing all roads and traveling up and down flights of stairs. Not all neighborhoods have concession options or public restrooms, so be sure to plan accordingly. Finally, Pittsburgh is a four-season city: stairs can become slippery when covered in snow, ice, rainwater or dry leaves and twigs. In the summer months, some hillsides become overgrown with invasive plants and block access to the stairs. Always wear sturdy footwear and be aware of the conditions around you. When visiting residential neighborhoods, respect all formal parking signs and informal "parking chairs" placed curbside. When in doubt about where to park, it never hurts to smile and politely ask.

It's also important to remember that Pittsburgh is a city of change. Our neighborhoods change, and the city steps that connect them also change.

The Middle Street city steps in East Allegheny and their commanding view of downtown.

Therefore, we can't guarantee that the descriptions we provide in each walk of conditions and surroundings will stay constant. In addition, stairs, sidewalks, bridges or roads may be closed by the Department of Public Works when deterioration occurs or repairs are scheduled. Bus routes may change, too, so it's always a good idea to check Pittsburgh Regional Transit (PRT) at rideprt.org to confirm bus routes and times. Also, while we have tried to verify the step count for every flight between Department of Public Works records and personal experience, accuracy is ultimately in the steps of the individual hiker.

Finally, never forget that your engagement is the most effective way to preserve these historical yet functional public stairways. Walk the steps and feel a linkage with the past while experiencing some of the best views in the area. Tell your friends and family about these walks and invite them along to explore with you. If a "boots on the ground" experience isn't available to you, consider following the tours virtually using Google Maps' Street View feature. Additionally, the Pittsburgh Mapping and Historical Site Viewer, a free online resource produced by the ESRI Mapping Center, allows users to view Pittsburgh's early street maps with an overlay of current locations. It's not exactly a time machine, but it does provide a view into Pittsburgh's development.

The city steps embody Pittsburgh's history, romance, intrigue and energy and are a remarkable symbol of this great city. Take pride in being a part of its long and storied narrative and advocate for the preservation of this unique municipal infrastructure.

HISTORY OF THE STEPS

*Note: From 1891 to 1911, the city's name was federally recognized as "Pittsburg,"
although use of the final* h *was retained during this period by the city government and other
local organizations. For ease of reading, the modern spelling is used throughout this section.*

Today, nearly all cities and towns, both large and small, benefit
from municipal infrastructure—the highways, bridges, roads and
sidewalks that have improved public travel over time. Pittsburgh, a city that
once boasted a population of nearly 700,000 people and large industrial
complexes, developed its transportation systems in the nineteenth and
twentieth centuries, along with a rising form of mass transit not found
in many other cities: public stairways. Locally referred to as "city steps,"
these stairs are a throwback to a very different time in history and a very
different Pittsburgh.

A City of Bridges

Few cities in the United States can match the dramatic changes to the
landscape that have occurred in Pittsburgh. Its natural environment
provides a striking panorama of flowing rivers, streams, steep bluffs and
deep valleys. Western Pennsylvania has never experienced major tectonic or
volcanic activity or any interruption to its local rock beds. In fact, the rugged
topography on this side of the state is largely a product of erosion from

The view from the 2nd Avenue steps. A barge passes the 10th Street Bridge as it travels the Monongahela River.

the outflow of glaciers over many thousands of years. Pittsburgh's elevation varies wildly, fluctuating 660 feet from highest to lowest points throughout the area, which makes it one of the hilliest cities in the United States.

The region's distinctive terrain, coupled with its iconic three rivers—the Allegheny, Ohio and Monongahela—and its rich geology, filled with coal and limestone deposits, played a substantial role in the city's early economic development and expansion.

At the same time, well-eroded plateaus and converging waterways posed a unique transportation problem for local residents. The solution is one of the city's defining characteristics. In addition to its many stairways, Pittsburgh is also known for its bridges, which play an integral role in the city's transit system. Following the Civil War, inclined railways and public stairways also came about to ferry a growing population throughout the steep hillside areas they called home.

FROM THE HILLS TO THE MILLS

Western Pennsylvania's natural resources were ideal for producing glass, iron and steel. For one hundred years, Pittsburgh's industries extracted sandstone for glass, coal for steel and oil for fuel. Mills and factories dominated the riverfronts on all sides of the city, as these waterways provided access to the world's shipping lanes, including the Mississippi River and the Gulf of Mexico, thereby easing the transport of supplies and goods.

B.F. Jones and James Laughlin developed the flatlands along the Monongahela River in Hazelwood and the South Side Flats in the mid-1800s. In time, the two would become business partners and form Jones & Laughlin Steel (also known as J&L Steel). Land on the east and west banks of the Ohio River became the industrial centers of Manchester and McKees Rocks, while the riverfront on the south bank of the Allegheny River became the Strip District. The area that was once a place for iron foundries and produce warehouses later transformed into a treasured destination for international grocery stores and retailers, boutique shops and independent merchants.

Pittsburgh's burgeoning industries depended on many manual and unskilled laborers and employed a massive influx of immigrants during the Gilded Age. The only affordable, inhabitable land for the working class was on the hilltops or, in some cases, along the hillsides. In an era before the near-universal adoption of motor vehicles, many residents needed to live within walking distance of their place of employment. As the area's population swelled in the late 1800s, so too did the need to transport workers "from the hills to the mills." Enter the city's steadily expanding network of public stairways.

While the steepness of the hillsides often deterred residential development, many Pittsburghers did construct homes on the slopes—

INCLINES: UPS AND DOWNS

Inclines—also known as funiculars or mountainside railways—can be found worldwide and are designed to transport people and cargo up and down steep slopes. Inclines operated throughout the hilliest of regions in Mount Washington, the Hill District and Troy Hill, among other neighborhoods.

Still in operation today, the Monongahela Incline, so named because it ran from the Monongahela Borough (today part of Pittsburgh's South Shore) to Mount Washington, was built in 1870 by engineer John J. Endres. Endres was assisted by his daughter, Caroline Endres Diescher, who was one of the first women engineers in the United States. The Monongahela Incline was the first passenger incline in Pittsburgh and the first in the United States and was added to the National Register of Historic Places in 1977.

The Mount Washington area, among the city's steepest, was a natural choice for Pittsburgh's first incline, and the Duquesne Incline, built in 1877, was the third to carry passengers. It continues to operate today thanks to local preservation efforts.

At the height of the incline's popularity, Pittsburgh had about twenty in operation, with several used for carrying coal. Of them all, only the Monongahela and Duquesne still exist. It just so happens that a set of city steps replaced the Fort Pitt Incline in 1919; they run along the old incline route from Second Avenue near the Armstrong Tunnel up to Bluff and the connecting pedestrian bridge to Duquesne University. The Knoxville Incline, in operation from

A 1933 view of the Mount Oliver Incline crossing Birmingham Street (Brosville Avenue) and Monastery Street near the St. Joseph Way city steps. *Pittsburgh City Photographer Collection, University of Pittsburgh.*

The Monongahela Incline passes over the P.J. McCardle Roadway connecting the South Shore to Mount Washington.

1890 to 1960, now has a foot trail and flight of city steps connecting Brosville Street in Allentown to Welsh Way and Fritz Street in the South Side Flats.

Inclines can be found worldwide and serve as both public transportation and tourist attraction. In addition to Pittsburgh, world-famous inclines are found in Lisbon, Portugal (Gloria Funicular); Los Angeles, California (Angels Flight); Haifa, Israel (The Carmelit); and Wiesbaden, Germany (Nerobergbahn), among others. A search of the internet reveals numerous "funicular fans" from around the world sharing locations, histories and travel tips for visiting this unique form of transportation.

they looked for buildable land close enough to their jobs but also above the dense layer of smoke billowing from the mills. To accommodate even more occupants, houses with two stories fronting a street often had an additional two or three stories descending along the back of the hillside. Many steep-slope residents reached their homes by ascending wooden stairways after a hard day's work. In the years before municipal water and sewer services, stairs also provided the only access to the community water supply and privy vaults, or outhouses, for many poor and working-class families.

EARLY TRANSPORTATION OPTIONS

Pittsburgh's first omnibuses—typically two horses guiding an enclosed carriage holding twelve to fifteen passengers along selected routes—appeared in the late 1830s. The most popular lines in the city were exclusive to the upper and middle classes, as the typical twelve-cent fare would have represented an occasional luxury for skilled laborers and far greater than the average unskilled laborer, who made between seventy-five cents and one dollar per day, could afford. In the late nineteenth century, the original omnibus routes expanded with horsecars, trams and trolleys drawn by horses over designated tracks. This popular system covered nearly fifty-six miles over various routes with fares priced to appeal to middle-class, white-collar workers.

Despite the growing adoption of trams and trolleys, most millworkers worked longer hours, were paid less and were more likely to walk. In addition, immigrant families frequently depended on the wages of all family members, which would have meant a multiplication of fare if employment was beyond walking distance. As the city moved into the first decade of the twentieth century, nearly half of the population lived within a mile of their place of employment. In neighborhoods close to large mills, more than 80 percent of residents walked to work. But by the late 1930s, ownership of automobiles had increased for many. A survey conducted by the Bureau of Business Research at the University of Pittsburgh revealed that, among chief income earners, a little more than half owned automobiles. However, that access was not uniform, and neighborhoods such as the Lower Hill District, a historic Black neighborhood with Italian and Jewish enclaves, had automobile ownership rates below 10 percent.

Edward Manning Bigelow:
"Father of the Parks" and So Much More

Edward Manning Bigelow is frequently referred to as the "father of Pittsburgh's parks" due to the transformation he brought to the city throughout his career as director of public works. During his thirty years working for the city, Bigelow's planning brought about significant improvements in boulevards, waterworks and parks, and he oversaw much of the neighborhood infrastructure improvements, such as paving, the transition from gas to electric streetlights and city steps and sidewalks.

Bigelow was born in Pittsburgh on November 6, 1850, and attended the University of Pittsburgh (then known as the Western University of Pennsylvania) as a civil engineering student. He was initially employed in the city's survey department and later served in the construction corps before being promoted to assistant city engineer, parks commissioner and city engineer. In 1888, he was elected through a city council vote to become the first chief of the Department of Public Works.

While Bigelow held three terms of office, enacting his plans for Pittsburgh created significant controversy and attention. His visions for park development made him unpopular with those who wanted to use the land for more building development. After being targeted by Republican state senator and construction magnate William Flinn for reported misappropriation of construction funds and contracts, Bigelow was removed from this position by the city council in 1901. However, he was restored to the position for a third term from 1903 to 1906. He continued implementing his system of grand boulevards, including Beechwood, Grant and Washington Boulevards, which were designed and constructed to connect Pittsburgh's urban parks.

Due to Bigelow's efforts in the 1880s, Mary Schenley donated her land to the city in what became Schenley Park. However, in July 1916, Bigelow publicly opposed the Mary Schenley Memorial Fountain installation, also known as

A *Song to Nature*, created in bronze and granite by sculptor Victor David Brenner.

Bigelow thought the idea of a fountain was a disgrace and that Pittsburgh should provide something more substantial to the memory of the donor of Schenley Park. Bigelow was quoted in the *Pittsburgh Post* on July 7, 1916, saying, "I believe a majority of the people of Pittsburgh will not approve it. The fountain, in a few years, will be an eyesore to the people and a disgrace to the city."

Bigelow did not live long enough to see how wrong his predictions about the public sentiment toward *A Song to Nature* would become. He died on December 6, 1916, a week after being reappointed by Mayor Armstrong for a fourth term as public works director. On December 19, 1916, Grant Boulevard was posthumously named Bigelow Boulevard in memory of him.

Today, Bigelow Boulevard is a three-mile thoroughfare connecting downtown Pittsburgh to Schenley Park in

In 1919, Grant Boulevard (Bigelow Boulevard) featured wooden city steps leading to Crescent Street. The 17th Street Incline appears in the distance.

Oakland. In December 2020, the section nearest the park was updated through collaborative efforts between the Department of Mobility and Infrastructure, the Department of Public Works, the University of Pittsburgh and community groups. The area now features expanded and enhanced sidewalks and crosswalk lanes, traffic calming and bus stop improvements and upgraded bike lanes, thus ensuring safe public access to Schenley Plaza and Park, the Carnegie Public Library, Natural History Museum and Museum of Art, as well as the many businesses and amenities throughout Central Oakland. The terminus of Bigelow Boulevard is at the intersection of Forbes Avenue and Schenley Drive, a short journey to where *A Song to Nature* continues to capture the attention of passersby.

City Steps 1.0:
Building Pittsburgh's First Public Stairways

Before automobiles and paved roads, the first city steps were made of wood. By 1870, the Road Committee—the city government entity responsible for implementing approved petitions for new steps and repairs—authorized the construction of three flights of steps for Boyd's Hill (the Bluff neighborhood, also known as Uptown), all connecting the hilltop to Second Avenue down below. Three years later, construction of stairs in the South Side Slopes began in earnest, including flights connecting Washington and Josephine Streets (Park Alley), Josephine and Sheridan Streets (Oak Alley), Yard Alley (now known as Yard Way, one of the longest flights in the city) and Ninth Street and Eighth Street to Brownsville Avenue. While the South Side Slopes dominated building efforts, stairs appeared in the Hill District connecting Somers Street and Wylie Avenue in 1875. Nearly every year following brought one or more city ordinances authorizing the construction or repair of city steps.

Another city government entity, the Committee on Gas Lighting, erected beautiful public lamps in the later quarter of the nineteenth century that illuminated a rapidly expanding landscape. The first gas lamp lit up a flight of city steps at Boyd Street on the corner of Linden (Lomond) Street in what is now Duquesne University in the Bluff neighborhood, and soon more followed all over the city.

Indian Trail steps looking toward Carson Street and downtown, showing the track for the Duquesne Incline. *Pittsburgh City Photographer Collection, University of Pittsburgh.*

Although the wooden city steps of the 1800s are long gone, one set, in particular, is worth revisiting. The Indian Trail steps were an impressive feat of engineering constructed during Pittsburgh's industrial beginnings. Built in the early 1900s, this flight ran along the side of Coal Hill (Mount Washington) from Carson Street, near the Duquesne Incline, to a point near the intersection of Shaler Street and Grandview Avenue in the Duquesne Heights neighborhood. This flight was more than one mile long with one thousand steps. Before these stairs were built, workers traveled a footpath that appeared on maps dating back to 1763 (a path likely used generations prior by Native Americans in the region). Despite the nearby Duquesne Incline, workers usually walked to avoid paying the fare. This flight was demolished in 1935.

The contours of Pittsburgh's topography made grid-style street patterns impossible. Aside from major corridors, nineteenth-century streets were often narrow with unusually steep grades, which affected accessibility. For those residing in isolated communities such as Skunk Hollow (below Bloomfield, on what is now the Busway) and Basso La Vallone in the valley below Larimer, wooden steps were the only way to connect residents of the hollow to those in the main neighborhoods above.

Not until the first quarter of the twentieth century would the city launch a concerted effort to widen streets, reduce steep grades, improve boardwalks and sidewalks and repair and build city steps.

BEYOND THE GILDED AGE

Public stairways began appearing later in the nineteenth century, and by 1937, more than thirteen miles of steps were available to pedestrians. By the start of the twentieth century, Pittsburgh's population had quadrupled to 322,000. As thousands of immigrants poured into Western Pennsylvania, the city increased its efforts to build and maintain municipal infrastructure. These efforts marked a clear turn in societal expectations and the role of local government. In 1920, the city budget allotted a little more than $74,000 for the construction of sidewalks and steps—roughly $1 million today—an amount that would increase throughout the decade.

Throughout the first two decades of the twentieth century, city steps were constructed in places such as Ohio River Boulevard on the North Side, Saw Mill Run Boulevard in the South Hills, Lilac Street in Squirrel

Hill South, Bigelow Street in Hazelwood, Irwin Ave in Perry South, Island Avenue in California Kirkbride, Woodville Avenue in West End Village, Broadway Avenue in Beechview and dozens more locations throughout the city. Due in part to the highway construction that would take place in the latter half of the twentieth century, some of these flights were demolished, but not all. Many would be replaced by the city in the post–World War II years and are still with us today.

HAZARDOUS TERRAIN

The well-to-do people drive to work. The medium people go on street cars and "inclines"—that's what they call them cable cars. And the poor people walk up the steps.
> —*Ernie Pyle,* Pittsburgh Press, *April 21, 1937*

As the 1920s came to a close, Pittsburgh found itself on the receiving end of claims brought against it by those unfortunate enough to be injured while walking up or down its steps. During the summer of 1929, Annie Valentine was awarded $500 (about $9,000 today) for the injuries she sustained while falling on the wooden steps between Butler and Baker Streets in the East End neighborhood of Morningside. These stairs were most likely an extension of Morningside Avenue, which in 1923 continued down to Butler near Mrs. Valentine's home address.

While the steps posed a very real threat of physical injury to unsuspecting travelers, they also attracted criminal activity, which often prompted public nuisance complaints. These ranged from serious crimes such as physical assaults, muggings and purse snatchings to verbal harassment, public drinking and loitering.

One of the most chilling and strange crimes on the city steps was the unsolved "Robot Gun" case reported in 1935. Colletta Madl, an eighteen-year-old neighborhood resident, was shot in the leg by an automatic weapon on the Doak Way stairs off Oakdale Street in Brighton Heights. The homemade gun was rigged to fire when a person stepped on the stairs. Fortunately, a passerby heard the shot and rushed Madl to a hospital. Police also discovered what the *Pittsburgh Press* referred to as a "maniac's warning," hinting at the possibility of additional devices hidden throughout the city. While the incident caused fear among residents, no other devices were

The poor condition of Beechview's Coast Avenue city steps was documented in 1936. Concrete stairs would be built in 1952. *Pittsburgh City Photographer Collection, University of Pittsburgh.*

found, and Colletta recovered from her injuries. The perpetrator behind the device was never found, and the wooden steps were reconstructed in concrete after World War II.

THE CYCLE OF REPAIRS BEGINS

As public stairways proliferated throughout the city, so did complaints about broken wooden steps and missing railings. While wooden stairs were convenient for pedestrians and relatively inexpensive to build, they had their shortcomings. In the days before pressure-treated lumber, Pittsburgh's weather quickly compromised their integrity. Heavily trafficked, they were also prone to accidental and intentional damage.

The *Pittsburgh Press* routinely published stories and opinion pieces about dangerous conditions on the steps. But residents did more than air their grievances in local newspapers. As early as 1927, the general public undertook definitive action by generating petitions and gathering signatures to demand specific repairs to flights of steps. Residents presented their petitions to district ward bosses and to the city council.

Perhaps because of this groundswell of community advocacy, Department of Public Works director Edward Lang rejected the city council's request for more wooden stairs leading from the Murray Avenue Bridge to Beechwood Boulevard in Greenfield. With durability in mind as one of the planning priorities, Lang requested $8,000 to use concrete instead, a viable solution to the never-ending repair cycle.

The economic downturn of the Great Depression in 1929 halted the city's plans to replace frail wooden stairs with concrete. Fortunately, however, help arrived in 1931 in the form of federal funding for the unemployed. As reported in the *Pittsburgh Press* that March, President Herbert Hoover approved the city's plan to employ four thousand laborers on thirty-six public works projects throughout the city, which included building concrete city steps, retaining walls, playgrounds and parks.

On November 29, 1936, the *Pittsburgh Press* reported that Works Progress Administration (WPA) crews had officially begun repairing more than three hundred flights of city steps. Designed to provide employment through infrastructure upgrades during the Great Depression, the WPA was created by President Franklin Delano Roosevelt in 1935 and lasted for eight years. Pittsburgh received $1 million, with $700,000 allocated for wages. Flights

The "Ways" of Pittsburgh

Whether you're out walking or looking at a map, you'll soon discover that many flights of city steps begin, end or intersect with a "way." A quick exploration, either on-site or through Google Street View, often reveals a narrow paved or semi-paved path that's a bit wider than a dump truck. When traveling down a way, a walker will likely see the backyards of houses, rear walls or loading docks for businesses, detached garages and an assortment of empty lots. In other cities, these areas are often called "alleys." At one time, they were referred to as such in Pittsburgh. However, on November 16, 1914, the city council passed an ordinance "[c]hanging the name 'alley' on every thoroughfare in the City of Pittsburgh to 'way.'" This change was thought to be more appealing to current residents and new arrivals because alleys were associated with crime and vice, whereas a way sounded more refined, pleasant and perhaps even a bit pastoral. This sentiment couldn't be more accurate than on Romance Way in the South Hills neighborhood of Carrick.

The Highnote Way city steps near Romance Way in Carrick.

Left: The Murray Avenue Bridge city steps connecting Greenfield to Squirrel Hill were fully reconstructed in 2023.

Below: The Oakley Street city steps in the South Side Slopes neighborhood after being reconstructed in concrete (1930). *Pittsburgh City Photographer Collection, University of Pittsburgh.*

Opposite: In the late 1920s, new concrete city steps were constructed alongside old wooden ones near the California Avenue Bridge on Pittsburgh's North Side. *Pittsburgh City Photographer Collection, University of Pittsburgh.*

considered "high traffic" justified concrete construction; for all others, wood would have to do. The list of steps needing attention included nearly one hundred flights on the North Side, forty-four in the far East End neighborhoods, sixteen in the Lawrenceville area, sixty-six in Oakland and Greenfield and ninety-five in the Mount Washington District encompassing both West End and South Hills neighborhoods.

Funding for public works flowed into Pittsburgh throughout the 1930s, but wooden stairways continued to prove hazardous both for residents and the city government. By 1935, lawsuits against the city in response to injuries suffered on the steps had become so frequent that Attorney General Charles J. Margiotti asked the Pennsylvania state legislature to pass a bill preventing fraudulent litigation and falsified injuries and claims. A subsequent investigation by the Allegheny County Bar Association uncovered evidence of lawyers colluding with doctors when prosecuting accident and damage cases. Despite these findings, injury claims on public property continued to make headlines. By 1938, the city's Law Department had come under scrutiny for paying out settlements upward of $375,000. Soon after, newspaper coverage shifted to exposing fraudulent claims, and the Department of Public Works started barricading flights it deemed unsafe. While victims of faulty steps continued to file legitimate injury claims, legal hearings and final judgments were likely to drag on for months or even years.

CITY STEPS 2.0:
PITTSBURGH PUBLIC STAIRWAYS IN THE POSTWAR ERA

Although World War II did not come to an end until September 1945, in May 1944, Pittsburgh City Council members, together with Mayor Cornelius D. Scully, proactively started planning postwar infrastructure updates. As reported in the *Pittsburgh Press* on May 10, 1944, a proposed ordinance to authorize a $15 million bond issue for infrastructure improvements was approved for a special election. Plans included remediating the floodplain along Saw Mill Run, replacing wooden city steps with concrete, updating sewers city wide, improving blighted housing in East Liberty and addressing other needs in individual neighborhoods. The bond initiative was designed for long-delayed repairs but also intended to provide employment opportunities for returning service members. The council members agreed to generate an extensive list of warranted repairs, and by July 18, 1945, they had finalized an official list of eighty-nine city steps plagued by structural issues and closed to the public. While the list represented all areas of the city, the North Side and the Oakland–Hazelwood corridor took center stage.

The planning that occurred throughout 1944–45 resulted in the beginning of a new era for Pittsburgh: one of concrete steps. In advance of the bond issue, the city council and Mayor Scully succeeded in including $100,000 for steps repair in the city's 1946 budget. The proposed list included eight flights throughout the city: Bell Isle Avenue, Wing Way, Rising Main, 56th Street at Carnegie and Duncan, Inglenook Place at Sickles and Oakwood, Tullymet Street, Alexander Street from Independence to Greenleaf and Lauer Way between Arlington Avenue and South 8th Street.

However, the public's patience was wearing thin, and the city couldn't make improvements fast enough, even with bond funding on the horizon. The state of the stairs was such a hot-button issue that throughout 1945, Republican mayoral candidate Robert N. Waddell, a popular football coach of Carnegie Tech (now Carnegie Mellon University), garnered consistent publicity for visiting neighborhoods with dangerous city steps and commiserating with frustrated residents. Waddell ran a competitive race but ultimately lost to State Democratic Party chairman and longtime Pittsburgh political influencer David Lawrence.

In 1946, the poor conditions of the city steps and the danger they posed to children were cited in protests against the board of education. The board had closed a Troy Hill K–8 public school and reassigned 258 students to schools in Spring Garden and East Allegheny, requiring longer walks.

The Inglenook Place city steps travel through a now-empty hillside in the East Hills neighborhood.

Left: The Lauer Way city steps connect Allentown to the westernmost section of the South Side Slopes.

Below: The St. Thomas Street raised sidewalk steps in the South Side Slopes.

Editorials criticizing elected officials and public works employees appeared in the *Pittsburgh Press*, including one written by Mary Draganjac, a young girl residing on Graib Street in Fineview who pleaded for assistance restoring her neighborhood's connection to Howard Street.

Fortunately, improvements would soon come to Fineview and many other neighborhoods throughout the city. For the next three years, Pittsburgh allocated $1,150,000 to remediate city steps previously closed and others in need of repairs.

Many newly constructed concrete stairs were impressive in size, visually striking against the hillsides and prominently displayed in photographs and stories in the local press. An article in the *Pittsburgh Press* from September 27, 1948, about the St. Thomas Street steps on the South Side Slopes detailed various construction methods, including slightly sloped treads to prevent ice buildup and a rough exterior surface for improved traction.

The new concrete stairs at McCandless and Stanton Avenues in Stanton Heights also made the news for a very different reason: stolen jewelry and bonds valued at $10,000 had been hidden underneath them. Local authorities apprehended Charles Sands, twenty-one, and Arthur Astridge, twenty-four, who confessed to the crime.

On Your Mark, Start the Clock

The city continued to allocate funds for sidewalks and stairs after the big infrastructure push of the late 1940s and early '50s, but with considerably reduced annual budgets. From 1958 through 1966, the budgeted allocation stayed at a steady $43,500, rising slightly from 1967 through 1970 but returning to the $43,500 baseline in 1971.

The average lifespan for a set of concrete stairs ranges between fifty and seventy-five years. While wooden stairs remained active, they had been reduced to serving mainly lower-traffic residential areas. In local press reports throughout the 1950s, the city steps transitioned from a heated headline news topic to the occasional public interest piece. In the June 15, 1956 edition of the *Pittsburgh Post-Gazette*, public works director James S. Devlin and city structural engineer John K. Roth were interviewed about the seventeen miles of concrete stairs constructed since 1946. Many of the flights discussed in the report are still familiar today, such as Sterling Street in the South Side Slopes and Rising Main in Fineview. Others, such as

THE DOCUMENTARY PHOTOGRAPHY
OF CHARLES "TEENIE" HARRIS

Charles "Teenie" Harris (1908–1998) was born and raised in Pittsburgh's Hill District neighborhood. He expressed curiosity about photography early in life, and his grandfather, who shared his interest, fueled this passion.

Throughout his life, Harris seriously pursued photography as his livelihood, and despite being self-taught, he secured and held the position as the preeminent photographer for the *Pittsburgh Courier*—a leading African American news publication—for more than forty years. Due to the swift

Opposite: Teenie Harris's photo of men constructing concrete stairs. These may be the city steps connecting Centre Avenue to Brackenridge Street in the Hill District. *Charles "Teenie" Harris Archive, Carnegie Museum of Art.*

This page: Teenie Harris's 1949 photo of Eleanor Street in the South Side Slopes compared to the same location in 2023. *Charles "Teenie" Harris Archive, Carnegie Museum of Art.*

and precise way Harris captured subjects, Mayor David L. Lawrence gave him the nickname "One Shot Harris."

During the 1940s, Harris's photojournalism for the *Courier* frequently captured injury and crime scenes associated with the city steps. These stark images highlighted the dilapidated and unsafe conditions many residents experienced throughout the World War II years and the infrastructure building boom in the years immediately following the war's end.

In addition to his work for the *Courier*, Harris also photographed scenes from weddings, sporting events and entertainment throughout Homewood and the Hill District's famous jazz bars and concert venues. His images of ordinary people and neighborhood life capture the Black community's struggles and resilience, showing a city and an era teeming with energy, culture, friendship and family.

In 2001, the Carnegie Museum of Art purchased Harris's collection of eighty thousand negatives from his estate. Thanks to the generous support of the National Endowment for the Humanities, the Carnegie digitized nearly sixty thousand of Harris's negatives from the 1930s to the 1980s and made them viewable online.

Through outreach efforts, lectures and special events, the museum has asked for assistance in identifying the people, places and events in the images. In 2020, a dedicated permanent exhibition of Harris's photographs was installed, titled *In Sharp Focus: Charles "Teenie" Harris*.

Dornbush Street in Homewood and Woodville Avenue in the Chicken Hill section of the West End Village, have faded into obscurity.

Unfortunately, not everyone benefited from a brand-new concrete flight during Pittsburgh's big building boom. On July 19, 1965, Mount Washington residents on Griffin Street appeared in the *Pittsburgh Press* to lobby for concrete steps to replace the wooden flight at Dewitt Street connecting their neighborhood to Southern Avenue. According to the article, volunteers tracked daily usage of the steps by nearly two hundred pedestrians, who likely made up many of the signatures on a petition delivered to the city council. Despite these efforts, it would take nearly a decade before the construction of concrete stairs was completed.

The current-day Dewitt Street city steps viewed from Southern Avenue in Mount Washington.

Starting with the 1960 U.S. Census, Pittsburgh's consistent population growth ended, and the era of significant depopulation began. While residents left the area and industry shifted, the concrete and steel stairways remained. While the city council approved an $80,000 ordinance to repair retaining walls and city steps in 1984, public support for this transportation method had cooled significantly. In 1985, residents in the South Hills neighborhood of Lincoln Place brought forth a petition requesting a public hearing to remove the city steps connecting Glenhurst Road to Lougean Street, citing ongoing repairs, litter and claims of public disturbance. The hearing did not rule in favor of the neighborhood, likely given the cost of demolishing the concrete structure. To this day, stairs connect the two streets.

CITY STEPS 3.0:
NEW CENTURY, NEW PUBLIC STAIRWAYS

The city emerged into the new century after decades of being unable to provide any attention to the steps beyond making essential repairs to a few high-traffic flights. Many of the concrete city steps built during the postwar infrastructure boom were reaching their "expiration date" in 2003, the same year Pittsburgh went into state financial oversight, otherwise known as Act 47, because of longstanding financial problems. Cuts to jobs and services and caps placed on spending prevented the city from filing for bankruptcy. It wouldn't be until the city budgets of 2016 and 2017 that funds for the steps would again hit the six-figure mark at $385,000 and $575,000, respectively.

In 2014, newly elected mayor Bill Peduto tasked city planners with conducting a citywide steps assessment to collect data and photos documenting each staircase's location and condition. Given the large number of staircases spread over Pittsburgh's fifty-five square miles, the city set up a call for volunteers. Pittsburghers' love for their steps was abundantly clear when three hundred people signed up within the first day to help. These data collection efforts resulted in the region's first online city steps map, providing residents, visitors and city employees with immediate, accurate information about stair location and conditions.

In 2017, the newly established Department of Mobility and Infrastructure (DOMI) launched the City Steps Plan, a citywide effort to prioritize

In 2018, the Joncaire Street city steps in Central Oakland were rebuilt with a lower profile for easier climbing, closed railings and enhanced lighting.

This page: A curious intersection of city steps and catwalk-style platforms at South Oakland's Frazier and Romeo Streets.

which sets of stairs needed repair or rebuilding. Through community meetings, field visits and survey collections, DOMI analyzed quantitative and qualitative data to determine what flights were critical for accessing public transportation, schools, grocery stores, hospitals, libraries, houses of worship and main street business corridors. The plan helped systematize and prioritize the upkeep of city steps in areas with the greatest need.

The first rebuilding project generated from the City Steps Plan was the Joncaire Street stairs in Central Oakland, which opened in October 2018. An $800,000 grant from the Pennsylvania Department of Transportation funded the work. The new stairs had a lower profile for easier climbing, closed railings and enhanced lighting for safety, an embedded bike track (or "runnel") to allow cyclists to walk their bikes along the stairs and stormwater and invasive species management for the hillside.

Several newspapers, television and radio programs and social media platforms covered the grand opening of the Joncaire Street stairs. While construction styles and communications media had changed considerably since the late 1940s, local interest in the city steps was as enthusiastic as ever.

In 2018, Pittsburgh regained its fiscal independence and is no longer considered a financially distressed municipality. In the years since the opening of Joncaire, DOMI and the Department of Public Works have worked with elected officials and communities to rebuild and repair many freestanding city steps and sidewalk steps around Pittsburgh. With more than nine hundred flights of public stairs in neighborhoods, parks and playgrounds, the to-do list is extensive, but the City Steps Plan, renamed the Pittsburgh Citywide Steps Assessment in 2022, provides a framework for prioritizing where the work is needed most. In addition to funds allocated through annual city budgets, securing state and federal infrastructure grants remains critical. While bridge repair and replacement has dedicated state and federal funding streams, work on city steps does not—the work is dependent on the city's budget. As a result, outside funding has been instrumental in large-scale rebuilding projects that cost hundreds of thousands of dollars in materials and labor.

In 2021–22, the city allocated nearly $4 million of federal funds received through the American Rescue Act to repair and replace a handful of city steps, including the Frazier Street stairs in South Oakland, Downing Street to Herron Avenue in Polish Hill and McCandless Street to Stanton Avenue in Lawrenceville. The following year, the federal government, with support from former U.S. Representative Mike Doyle and Senator Bob Casey, awarded the city $7 million explicitly for repairing the city steps. But with

Public Stairs Outside Pittsburgh

Pittsburgh has the largest number of public stairways compared to any other city in the United States, but the stairs aren't limited to the city proper. Throughout Western Pennsylvania, in former industrial towns featuring hilly terrain, public stairs can still be found connecting residential areas to spaces that once housed mills and factories. Depending on the municipality and the availability of archived historical maps, stairs may be easy to locate (as in Pittsburgh) or more of a challenge. Fortunately, local resources—such as libraries, historical associations and related museums—combined with online resources, including digitized maps and discussion forums, can yield stories and promising leads. Perhaps not so surprising is the luck that can often be encountered simply by walking through a neighborhood and politely asking residents about the location of stairs nearby.

Of course, Pittsburgh is not alone as a city with wild elevation changes, and there are many other cities with hundreds of public stairways for exercise enthusiasts and historians to explore. Some cities, such as Seattle, Los Angeles and San Francisco, have several guidebooks (not unlike this one) offering a collection of walking routes intermingled with regional information designed for visitors and residents alike. For other areas with many public stairs, such as Cincinnati and the Bronx in New York City, if a guidebook is unavailable, often the city government or local enthusiasts have assembled multiple online resources. Whether you prefer a book in your hand or digital maps for your smartphone, chances are the information is out there. If it isn't, your own investigations could prove fruitful for you and your fellow urban hikers and armchair explorers.

a staggering per-flight repair cost averaging between $500,000 and $1.5 million, the number of flights improved is limited. Once locations are chosen for repair and maintenance, two to three years can pass before construction begins because of legal agreements, federal regulations, proposed designs from engineering firms and contractor bidding.

While the days of residents walking from the "hills to the mills" have disappeared into the past, the stairs have gained in popularity. They offer an alternate form of public transportation to those seeking it out and connect hikers and cyclists to a growing network of designated trails and bike lanes throughout the city. Following in the footsteps of the engineers and laborers who constructed the city steps for the last 150 years, each building project remains unique. No two flights of Pittsburgh city steps will ever be identical, thanks to the unruly terrain.

AN INSIDE VIEW
OF PUBLIC WORKS

Patrick Hassett's tenure with the City of Pittsburgh spans several decades. Before his retirement in 2017, he worked as the principal transportation planner, assistant director of city planning and deputy director of the Department of Public Works (DPW). As deputy director, he oversaw the Bureau of Transportation and Engineering, a division responsible for programming all engineering and construction of the city's infrastructure projects, including its roadways, bridges, parks and trails and public stairways.

In 2022, he sat down for an interview to address ongoing curiosities about the maintenance, construction and care of the city steps. This Q&A outlines the role of Pittsburgh's DPW and how residents can advocate for infrastructure that meets their neighborhoods' needs.

The city has over nine hundred flights of public stairways. What role does the DPW play in their construction and maintenance?

There are four areas administered by the DPW: closures, general repairs, subcontracts for larger repairs and collaborations with other city departments when stairs need to be rebuilt, such as the Department of Mobility and Infrastructure (DOMI) and city council.

As city steps are part of the public right-of-way, DPW and DOMI are the authorities when they need to be closed because of unsafe conditions. The department's engineers are responsible for making the recommendation, and it's a serious process because professional licensing is involved.

With so many old flights of stairs, who should residents approach with concerns about unsafe conditions or closures for safety reasons?

Community advocacy is the best way to address conditions that are unfavorable to residents. Unless the public alerts DPW and their city councilor, the issue may never be resolved. Reporting the issue to 311, which serves as the eyes and ears of the neighborhood, is the first step. Calling and communicating with the city council representative and public works, and even writing letters and gathering petitions from other neighborhood residents, makes a difference. However, it's important to note that the process takes time, and in Pittsburgh, the needs for repairs greatly outweigh the resources available. In cases where we didn't have the resources, we still would need to close the steps or, more accurately, "mothball" them until the resources became available.

Tell us about Pittsburgh's City Steps Plan, created during your time with DPW.

The plan was created by DOMI, and it addresses how to determine which flights are most essential to residents and, by extension, how to allocate funding and prioritize repairs, which is where DPW enters the picture. The City Steps Plan ranked all flights of stairs based on community usage feedback, access to public transportation and ridership levels and the degree of mobility hardship placed on residents if stair access was restricted. A group of interns working for the city spent the summer of 2014 consulting Bob Regan's books, first to verify his data and then to report on the current condition of the city steps. Their efforts helped to create publicly available maps and databases of the stairs' locations and how important they are to residents for accessing transportation.

What contributes most to wear and tear on city steps?

Because all engineering and construction decisions came out of the Department, I relied on my professional staff to provide me with their assessments and recommendations. In the case of steps, I was fortunate to have Construction Supervisor Tom Joyce, [given] his experience working with concrete contractors.

Concrete stairs can maintain their structural integrity for many decades, but they don't last forever. Many of the flights around the city were constructed in the years following World War II, which makes them over seventy years old. So, age is a factor. But so is erosion as concrete slowly deteriorates over time as water can penetrate and have corrosive effects. As a four-season city, Pittsburgh gets plenty of rain and snow, and that causes natural erosion, as well as landslides. Additionally, because of the aging underground infrastructure of the Pittsburgh Water and Sewer Authority (PWSA), erosion can happen when old pipes begin to leak. Like the maintenance constraints the city has experienced with its above-ground infrastructure, PWSA has experienced the same. Seepage weakens the ground's stability, which also contributes to erosion.

Some stairs are also very susceptible to vehicle damage. For example, a car or truck can easily damage metal railings and wooden flights. During my tenure, engineering plans and design work developed preemptive measures to address these potential problems. We also developed improved solutions for essential components, such as railings, that reduce moisture build-up and can stop cement from cracking and weakening. These changes are not generally noticeable to the average person, but they're part of the engineering and construction process.

Speaking of railings, why is there no uniform color choice?

Sometimes a railing color was decided based on community input. In other cases, DPW used a very illuminating color intentionally, like the bright green on Rialto Street in Troy Hill, to improve visibility for motorists and pedestrians. Other colors were used well before my time working for the city, but I imagine the crews used whatever was available in the shop.

In the past few years, the city has started rebuilding older, historic flights, such as Joncaire Street in Oakland and Vista Street in East Allegheny. Has the design process changed over the years?

Not too much, to my knowledge. The design process always starts with, "How will I build in this location?" Every site is unique and requires different solutions. DPW works with other city departments, city councilors and often the community to determine the needs of a neighborhood. Labor, logistics and supplies and contractor performance were constant factors for every job. Larger jobs also needed their own bond issued, as the cost can run into the high six figures.

The view from the Lowrie Street Bridge of Troy Hill's infamously steep Rialto Street and sidewalk steps.

Describe some of the strides Pittsburgh has made in developing infrastructure to better support cyclists.

Moving toward greener transportation was a key focus of Mayor Bill Peduto's administration, and given the city's topographical challenges, placing bike runnels on city steps was seen as an effective way to connect trails and communities. The idea was that cyclists could walk their bikes up or down the stairs by placing the wheels on the track. We placed early runnels in Oakland on Louisa Street and Greenfield Avenue to connect cyclists to established bike paths in Four Mile Run, Schenley Park and Oakland. Later, the construction options for runnels changed and grew as we received feedback and learned what worked and what needed improvement. The bike runnel on Rialto Street may be the longest in the country.

Was there a time when DPW cleared all the snow from the city steps, or is that an urban myth?

I can't speak for the entire history of DPW, but while clearing sidewalks has long been the property owner's responsibility, the stairs have been

Cyclists appreciate bike runnels on city steps, but the city's first attempt on Joncaire Street in Oakland fell short.

Snow-covered steps of Sunday and St. Ives Streets in the California-Kirkbride neighborhood.

DPW's domain. Back when Pittsburgh had a much larger population and more extensive public works staff, I'm sure there was a priority list of stairs that needed clearing, but shoveling all city steps in Pittsburgh would be equivalent to scaling Mount Everest. So, in recent decades, the city has counted on community organizations and residents to help clear their stairs.

SNOW PATROL

In 1945, the city mandated snow removal from sidewalks in front of residences and businesses, and failure to comply could result in a five-dollar fine or ten days in jail. However, the mandate had more bark than bite—both the head of public safety and public works admitted that significant snowfall or protracted periods of below-freezing temperatures would make the task all but impossible.

Merely mentioning the mandate in anticipation of the season's first snowfall would automatically trigger a flurry of newspaper article comments and letters to the editor. Residents voiced their disapproval of the city's failure to clear city-owned sidewalks and steps and lambasted the snowy walkways left untended by elected officials and city employees.

During the late 1960s, the city enlisted two hundred Neighborhood Youth Corps members to assist DPW with clearing city steps, crosswalks and sidewalks during winter snowstorms. A federally funded dropout prevention program, the NYC aimed to provide paid employment to economically disadvantaged youths and improve high school graduation rates. Pennsylvania was the fourth-largest participant in the program behind Illinois, California and New York. When federal funding ended, the city terminated the program.

While the snow removal ordinance remains on the books, enforcement has been spotty at best. To this day, residents continue to delight in pointing fingers and shovels at city officials who neglect to clear their own sidewalks.

Do you have a favorite flight of city steps?

Honestly, I can appreciate them all. They're all so impressive. But I always liked the Bohem Street stairs and the I-376 overpass in South Oakland and never agreed with the decision to remove them. The issue was more about aesthetics and the visual approach to the city and less about their safety or use. That's one I wish we were able to keep. They would have maintained direct and dedicated pedestrian access from Oakland to the riverfront.

REBUILDING CITY STEPS

In September 2022, the remediation of several sets of city steps on Copperfield Avenue in the South Hills neighborhood of Carrick began. Located on the southeastern edge of Pittsburgh, Carrick is bisected by the Brownsville Road commercial corridor, with Copperfield serving as a connector between Brownsville and Saw Mill Run Boulevard. In 1853, Dr. John H. O'Brien received permission from the U.S. Postal Service to establish a post office in the area, and he chose the name after his hometown, Carrick-on-Suir, in Ireland.

The western section of Carrick between Volunteers Field and Phillips Park dates to the early 1870s and was originally referred to as the Keeling Plan. By the early 1860s, Brownsville Road, which travels from Arlington Avenue near the South Side Slopes to South Park, was already a well-established travel route. Within twenty years, Copperfield Avenue (originally known as Center Avenue), along with five intersecting streets and several properties, took shape. In 1926, Carrick's residents voted to become part of the city of Pittsburgh. Once home to prominent mansions and wealthy families, the neighborhood still offers family-oriented residential areas interspersed with business districts.

Newett Street, which runs parallel to Copperfield one block away, is one of the steepest streets in Pittsburgh, with a 23 percent grade. Copperfield's grade is similar, but its many flights of city steps make it much easier to travel. The original sidewalk steps along Copperfield were constructed in 1949 with eight distinct sections starting near Plateau Street and traveling to Brownville Road for a quarter mile.

Left: The old red brick water channels that once lined Copperfield Avenue were congested with debris and litter.

Right: Removing the six-block-long flights of sidewalk stairs and replacing them with new construction took several months.

These original city steps were a mix of sidewalk steps, raised catwalks and standalone flights. Their construction was standard concrete, but to address water runoff, a brick-lined channel ran parallel to the road and emptied into the storm drain at the end of each block. This channel, which was about two feet across by one foot deep, required short concrete sidewalk bridges to connect the street to each house. While this engineering concept for moving stormwater may have seemed ideal at the time, today, such infrastructure is considered hazardous to pedestrians and vehicles and ineffective because the channel collects weeds and litter. Overland flooding occurs after excessive rainfall or snowmelt, and floodwaters can quickly submerge roads and sidewalks, causing hazards. The old brick water channels, clogged with leaves and other debris, were contributing to, rather than remediating, flooding concerns.

To address these issues, Pittsburgh's Department of Public Works faced a monumental task. A job of this size required working with an approved and experienced subcontractor for the demolition and rebuilding. The aggressive timeline for completion—before temperatures consistently dropped below freezing (approximately six weeks)—required a large crew and the equipment and skills to quickly execute all stages of the project. The subcontractor would be responsible for removing the old structures; preparing the surface; building the frameworks for the new stairs, sidewalks and retaining walls; pouring and finishing the concrete; installing curb cuts and access to each house or commercial building; and ensuring all sections of the street affected by the construction were repaved in asphalt.

From day one, the site takes on the qualities of an assembly line. All overgrown weeds and grass are cut back, gas lines are marked and security cones are placed along the street to prevent vehicles from parking. As many new sections will be constructed using the wide "jumpwalks" style (a sidewalk with individual steps placed at a greater distance from one another), all existing infrastructure, including the brick water channel, needs to be removed. As the work begins, a resident on the street comes out to ask where they should move their car and smiles when they hear new stairs and sidewalks are on the way.

Excavating equipment arrives, and the sidewalks and stairs from Brownsville Road to Dellrose Street are removed. In the days and weeks ahead, the crew will work down each of the four blocks on the southern side of Copperfield and then proceed in the same direction on the northern side of the street.

Once the first block is cleared of all concrete and brick, the area for the new construction is measured, and the locations for the sidewalk stairs and the curb are delineated using stakes and twine. Using excavation equipment and manual labor, crews dig out the space to a depth of slightly more than two feet. This depth is needed to accommodate a layer of crushed stone that rests below the poured concrete and the city-required twenty-four-inch steel reinforcements that anchor the curb into the ground and to the sidewalk. Because of the project's accelerated pace, everyone is focused on the job at hand. But a few crew members stop to inspect an older brick sidewalk they unearthed that predates the 1949 construction. This layer of history will remain in the ground as fill, but for a few moments, they reminisce about artifacts found on past job sites: old bottles and cans, machinery and industrial parts and hand-cut stone slabs and curbs.

Using the established lines, wooden frames are built for the curbs and the sidewalks. Rebar is cut and fitted horizontally and vertically within each frame to provide stability. Footers are dug at regular intervals at a depth of an additional two feet to provide additional stability against gravity and hillside erosion. While one crew preps the first section for concrete, another begins excavating the second block of Copperfield between Dellrose Street

and Westmont Avenue on the southern side. This block poses an additional challenge, as it has elevated sidewalk stairs supported by a four-foot-high retaining wall of stone and an adjacent water channel. All must be removed and rebuilt to modern standards.

Once the wooden forms are completed on the first section, the curbs receive concrete first. About a third of the way through the pour, the operator is concerned about the truck's brakes (a legitimate worry on a street with such a steep grade), and he needs to make a hasty exit. This causes consternation for the crew since the concrete must be poured quickly and uniformly to ensure structural integrity. Fortunately, the situation resolves itself quickly, and the pour is finished. Once the curbs have hardened and the wooden forms removed, the concrete can be poured to create the sidewalks and stairs.

After the second concrete pour, work on the first block nears completion, and the excavation of the second block between Dellrose Street and Westmont Avenue ends. The crew moves down the street, and the process repeats itself in familiar ways: crushed stone is poured, curb and sidewalk lines are established, wooden frames are constructed and deliveries for concrete are arranged. But every construction is unique, and this block also needs a new retaining wall and crushed stone to help with water runoff and hillside erosion. As the crew foreman and project manager work out engineering details, the excavation operators get to work on the next block between Westmont Avenue and Oakhurst Street.

In this manner, the work proceeds from one block to the next and from one construction stage to the next, with regularity and a certain degree of predictability. As the project progresses, days turn into weeks, and the temperature begins to drop. The crew swaps T-shirts and bandanas for layered sweatshirts and wool hats as they chat about the unpredictability

Opposite: A newly constructed flight of city steps featuring a lower step profile emerges.

Above: Completed in early 2023, the new Copperfield Avenue city steps feature jumpwalks connecting flights of stairs and crushed stone for water runoff.

of the weather and the confines of a tight schedule. In addition to heavy equipment, crushed stone, lumber and construction supplies lining the streets, special blankets are laid over newly poured concrete to keep out dampness and fight against dropping ground temps. Weighted and waterproof tarps ensure that new curbs, sidewalks and stairs cure and develop properly without compromising strength and integrity.

As December approaches, the crews finish the final touches and turn the project back over to DPW, which will install the new handrails when warm weather returns. While the new construction appears wider than the original, it's a bit of an optical illusion. Because the brick water channel was removed, Copperfield Avenue is now a slightly wider street. The extra two feet will allow for easier deliveries, weekly trash pickup and parking in front of homes. While living in the middle of a construction zone for several weeks has its drawbacks, the new stairs and sidewalks enhance the appearance of the street and adjacent properties with significant upgrades to their overall functionality and ease of use.

PITTSBURGH'S NO. 1
CITY STEPS WALKER

Born in 1920, Anthony J. Mainiero, affectionately known as "Mr. Tony" to neighbors and friends, is a lifelong Pittsburgh resident with fond memories of the city steps. Born and raised in the East End neighborhood of Larimer, Tony loves sharing his childhood memories of buildings, churches, schools and steps that have long disappeared from the neighborhood. Listening to his descriptions of daily walks to school, running errands or hawking newspapers is a bit like traveling along a 1930s street map.

In the early twentieth century, Italians primarily resided in Larimer and East Liberty, the Hill District, Bloomfield and the Strip District. The largest Italian neighborhood, according to the 1920 U.S. census, was Larimer, with a total population of 6,061.

A residential neighborhood bordering the commercial corridor of East Liberty, Larimer is six miles from downtown Pittsburgh and one mile from the Allegheny River. Two valleys provide natural boundaries and divide Larimer from Highland Park to the west and Lincoln-Lemington-Belmar to the east. A railroad track to the south separates it from Shadyside. Because of these natural and man-made divisions, Larimer has three bridges to facilitate connections: the Larimer Avenue Bridge, Meadow Street bridge and the Lincoln Avenue Bridge. Owing to significant changes in elevation between the valley below and the plateau above, city steps were also necessary. While many of the stairways constructed in the 1920s and '30s have since been demolished or closed, Tony's childhood memories offer a glimpse into the daily life of Larimer residents a century ago.

Larimer's Lenora Street looking toward Chianti Way in 1930. A makeshift boardwalk to the valley below is visible.

Deep in the valley below the Meadow Street bridge, in the present location of Negley Run Boulevard, Chianti Way was a dirt street whose name reflected the Italian heritage of its residents. It was known as Basso La Vallone, or "down in the hollow," and residents in this valley community were frequently new immigrants of limited means who grew fruits and vegetables, planted grapes for wine, managed small vineyards and raised livestock in order to survive. "Even during Prohibition, you were allowed to make wine for your family," Tony explains, "and the hillside of Chianti Way was where the people grew the grapes for wine." City steps connected the upper part of Larimer to Chianti Way and the homes in Basso La Vallone. Flights were built at the end of Orphan Street to Burpee Street and Venus Way to Chianti Way. Another flight, constructed at Leonora Street and Pace Way, still exists but is closed because of structural decay. During Tony's childhood, the hillsides leading to Negley Run had streets and houses, a very different sight than what we experience today. Much of the area was filled in, and the houses were removed when the valley was regraded for Negley Run Boulevard during the 1950s.

Continuing south through the neighborhood toward Larimer Avenue, Tony proudly points out that the Larimer Avenue Bridge once held the record for the world's longest concrete reinforced span. Built between 1913 and 1917 as part of a city-initiated bridge-building program, the structure typifies the then-popular City Beautiful movement with its classical-style architectural details. Open spandrel arch bridges, such as the one for Larimer Avenue, were well suited to long-span crossings in need of a strong visual impact. Unfortunately, while the bridge is still used today, it has deteriorated, and its aesthetic qualities are significantly reduced by a black net designed to catch debris from falling onto Washington Boulevard below. Still, it remains an impressive structure and represents a significant engineering and design achievement.

In Tony's youth, the walk from the neighborhood to Washington Boulevard was easy thanks to several different flights of city steps, including the Thompson Street stairs at Paulson Avenue, a flight at Renfrew Street and Xenia Way and another at the entrance to the Lincoln Avenue Bridge.

A woman descends the Pace Street city steps in Larimer in 1929. These steps were later reconstructed in concrete but have since been abandoned. *Pittsburgh City Photographer Collection, University of Pittsburgh.*

The latter, which still exists today but is closed because of structural decay, was how Tony traveled to Westinghouse High School. This flight was also ideal for traveling to Silver Lake, a spring-fed pond popular for skating and swimming. "As time went by, the water got worse," Tony explains. "There were open sewers coming from Highland Park and Point Breeze, but we were kids, so we didn't mind it too much. We wanted a place to swim." By 1947, Silver Lake was closed to the public, and the pond was drained and filled. In its place, the 550-car Silver Lake Drive-In opened in 1949. Featuring in-car speakers, it was the only outdoor movie theater in the city of Pittsburgh. After the drive-in closed in 1968, the area transformed into an industrial park. A "Silver Lake Drive" street sign is the only remnant of days gone by.

Tony's former church and school, Our Lady Help of Christians, once stood a few blocks from Lincoln Avenue and the bridge. Construction began in 1898, and for nearly a century, the church was a community focal point. In 1947, after serving in World War II, Tony and his wife, Edi, whom he met while stationed in Italy, were married there. But decades of steady declines in membership forced Our Lady to close in 1992, and in 1995, the property was sold. After two decades of significant deterioration, the church and convent were condemned and fully demolished by 2021. However, the school building still stands today as the Urban Academy, a public charter school.

Tony and Edi settled into family life in Larimer, purchasing a home on Mayflower Street in 1949. Tony first worked as co-owner of a beer distributor and later joined the U.S. Postal Service as a letter carrier. For more than sixty years, he handled routes throughout Oakland and the Hill District, including the Terrace Village public housing complex and its 1,875 families. Located on Goat and Gazzam Hills, Terrace Village was one of the ten largest residential construction projects in the United States at the time. "The kids always looked forward to seeing me," he says. "They'd follow me around like I was the Pied Piper. I had that route for years and never had any trouble."

There were dozens of public stairs throughout these hilly neighborhoods, and Tony walked up and down most of them every day to deliver mail and packages. Despite the passage of time, he clearly remembers the poor conditions of the stairs built in the 1920s and '30s and the difficulty of driving the mail truck in the winter along the narrow and steep roads made of Belgian block. "I still remember those steep streets!" he recalls. "Chesterfield Road [near Carlow University] and Burrows Street. They were tough ones." Throughout the 1950s, USPS trucks still used 1930s-era metal frames with a wooden truck bed that held sacks of mail. Heavy chains were installed on the tires, and Tony wrapped his boots with burlap and rope to provide more traction while walking on the ice and snow.

He laughs when asked if public works routinely cleared snow from city steps in the winter. "What? No way," he exclaims. "I cleared the steps as I walked on them, but city workers didn't do that. They stayed covered in snow until the people who used them cleared them off." Many flights didn't have streetlights, making them extra treacherous in the early morning hours. "I had really good eyes back then," he reminisces. "I could see in the dark just fine, but you had to be careful."

Despite having a union, working ten- to fourteen-hour days without overtime was routine for mail carriers throughout the 1950s and '60s. In the '70s, that would change, but the hours were long, and there were always steep hills and city steps to navigate. Now 103, Tony no longer treks the steep steps of his postal service past but still reminisces about the many people who greeted his deliveries with a smile or a kind word. "I've always kept a positive attitude throughout my life," he shares. "It hasn't always been easy, but I've always made the best of it and tried to be the best person I could be. My glass is half full."

GUIDED WALKING TOURS

SPIRITED STAIR STEPPING ADVENTURES
AND URBAN HIKES

One of the most frequently asked questions about city steps is just how many exist, a question with no definitive answer. Bob Regan located and mapped 739 flights in his book *Pittsburgh Steps: The Story of the City's Public Stairways*, but in the years that followed, the Department of Public Works and CitiParks assumed shared responsibility for public stairways in the 163 parks and playgrounds throughout Pittsburgh. An additional 3,800 acres of parkland boosted the number of city steps on the official roster to more than 900. Also, in recent years the city has made a concerted effort to establish and map trails throughout the city's urban greenways. These efforts have produced many short flights of wooden stairs, typically embedded in the hillside, to improve trail access.

Pittsburgh also has many long flights of city steps that intersect with multiple cross streets. Flights like Yard Way, located in the South Side Slopes and one of the longest in the city, further complicate the stairway count. While the general public may count it as a single flight, with endpoints at Pius Street and St. Paul Street, the city's Department of Public Works divides it into seven distinct flights: Pius to Gregory Street, Gregory to Magdalene Street, Magdalene to Roscoe Street, Roscoe to Baldauf Street, Baldauf to Huron Street, Huron to Shamokin Street and Shamokin to St. Paul Street. By parsing longer flights, the city can handle repairs and record the public right of way with better efficiency.

While the paved portion of Virginia Avenue ends at Plymouth Street, wooden city steps connect to two additional roads in Duquesne Heights.

Regardless of how you calculate the total number, 344 flights of stairs are considered legal streets, many of which have specially designed street signs indicating that the street is a set of steps. The city also has nearly 75 flights of wooden stairs. While wood was the building material of choice for flights constructed in the late 1800s and early 1900s, that practice was primarily replaced by concrete and steel construction in the late 1940s. However, wood is still occasionally used because of its affordability and ease of construction and repair. Unfortunately, weather conditions and vehicles can easily damage wooden stairs, so it never hurts to exercise a little extra caution when traveling on one.

In the days before robust GPS systems, online maps commonly misdirected drivers to a flight of stairs. Fortunately, with those days behind us, mapping programs now offer directions based on transportation modality. For Pittsburgh pedestrians, that means walking directions will often include city steps. However, it's important to note that walking directions will likely not reroute around recently closed city steps and may not include all traversable flights of stairs.

For those intrepid urban hikers wanting a comprehensive list of all public stairways, the Western Pennsylvania Regional Data Center's City of Pittsburgh Steps (https://data.wprdc.org/dataset/city-steps) provides all the information needed to locate every flight. Just remember, the list is dynamic and actively used by the city, so check it regularly for updated information.

Ninety Neighborhoods and 900-Plus Flights of Stairs

Pittsburgh's many diverse neighborhoods are as fascinating as their city steps. The city recognizes ninety neighborhoods throughout its four regions (see https://gis.pittsburghpa.gov/pghneighborhoods for locations). With more than 900 flights of stairs scattered throughout hillsides, sidewalks, parks, trails and playgrounds, walkers never have to travel far to find them. The South Side and South Hills have the greatest number by far at 455.

The website for Pittsburgh's Department of Mobility and Infrastructure's City Steps Plan (https://pittsburghpa.gov/citysteps) offers an interactive map showing the location of all steps under the city's care.

GAPS IN THE RAILINGS

When traveling the stairs, it's common to come across openings in the railings that were intentionally designed and constructed. Many walkers don't give them a second thought, but it's important to look out for them. At one time, these openings led to a home or, in some cases, an entire alleyway (referred to as a "way" in Pittsburgh) leading to several homes. Today, many flights of city steps lead walkers through a nearly parklike experience over a hillside covered with greenery. Still, it's important to remember that one hundred years ago, the view from these stairs was quite different. The stairs were constructed to transport a city of nearly 700,000 people but now exist in a city of 300,000. Many structures have been demolished over the decades, and nature has reclaimed the land. When walking the stairs in winter, it's still possible in some places to see the foundations and retaining walls of the homes that once occupied these hillsides. In early spring, the hillsides offer a home to blooming lilacs, daffodils and crocus—a reminder that many years before, a garden was planted there.

On the South Side Slopes' St. Michael Street city steps, a gap in the railing leads to hillside houses.

NORTH, SOUTH, EAST AND WEST

Pittsburgh is divided into geographic quadrants: the North Side, East End, South Side and West End. A short description of each quadrant provides a bit of background for walking tours in each area.

North Side

The North Side, formerly known as Allegheny Ciy, is north and east of the Allegheny and Ohio Rivers. Incorporated into the city of Pittsburgh in 1907, this area has been well known to sports fans as the current location of Acrisure Stadium (formerly known as Heinz Field) and PNC Park. One of the hilliest sections of the city, the North Side boasts the highest point of elevation (1,379 feet), and yet it "only" has 232 sets of steps. A lack of steep topography abutting the riverbank, except for the area near Troy Hill, made it an ideal location for stockyards, tanneries and meatpacking plants, as well as the H.J. Heinz Company along the shore of the Allegheny River.

TOURS ON THE NORTH SIDE
"Troy Hill: Where Everything Old Is New Again"
"Spring Garden and Spring Hill–City View: Put Some Spring in
    Your Step"
"Fineview: Living Up to Its Name"
"The Six Longest Flights"
"Pittsburgh's Steepest Streets"
"Off-Road Adventures"

East End

The East End includes a diverse collection of neighborhoods with differing economic, ethnic, cultural and racial backgrounds. It's also home to one of the most international of all the city's neighborhoods: Oakland, home to the University of Pittsburgh and its associated hospitals, Carlow University, the Carnegie Museum of Art and Natural History and Pittsburgh's main branch of the Carnegie Public Library.

While the pedestrian bridge connecting South Oakland to Hazelwood was demolished, the remains of the Bohem Street city steps perch above I-376.

Perhaps the most spectacular set of steps in this area stretched from Second Avenue to Bohem Street, which used to cross the Parkway East, connecting a high point in the South Oakland neighborhood to the former J&L steel mill site. The steps and their pedestrian overpass were preserved when the parkway was built in the 1950s but have since been demolished. However, remnants can be seen on the hillside from the inbound lane and through Google Maps Street View's timeline feature (see "Resources" at the end of this book).

Tours in the East End
"Polish Hill and the Hill District's Upper Hill: A 360-Degree Pittsburgh
    Experience"
"Oakland: Panther Hollow Magic for Muggles and Marauders"
"Hazelwood: With Every Step, Another Story Unfolds"
"The Six Longest Flights"
"Pittsburgh's Steepest Streets"
"Off-Road Adventures"

South Side and South Hills

The South Side of Pittsburgh aligns with the Monongahela River. It consists of "the Flats," the neighborhood adjacent to the river; "the Slopes," the high ground beyond; and the "South Hills," a large collection of inland residential and commercial neighborhoods. This region of Pittsburgh is dense with city steps, many of which can be seen from across the river.

The South Side Slopes neighborhood typifies the early immigrant neighborhoods perched above the flat area occupied by the mills. Its houses, typically one room wide and several stories high, packed together on a hillside with many narrow walkways, provide a glimpse of an early millworkers' neighborhood. This region has many narrow winding streets and a dense network of city steps still in routine use.

The South Hills are connected to downtown Pittsburgh through a commuter train line called the T. With three distinct lines (Red, Blue and Silver), the T provides public transportation to many South Hills neighborhoods—including Beechview, Beltzhoover, Overbrook, Brookline and Carrick—and it continues outside of Pittsburgh to other communities in Allegheny County.

Tours in the South Side and South Hills
"South Side Slopes West: Stairways to Heaven"
"South Side Slopes East: Sterling Efforts Yield Rewards"
"Beechview and Beechwood: Traveling through Time"
"The Six Longest Flights"
"Pittsburgh's Steepest Streets"

West End

The West End is both a formally designated neighborhood (also referred to as West End Village) and the section of the city south of the Ohio River, which has its genesis at "The Point," or the combined terminus of the Allegheny and Monongahela Rivers. Often called the best-kept secret in Pittsburgh, this area offers some of the finest views of the city, both from bluffs above the river and from the West End Bridge. The West End, like many areas in the southern part of the city, bustled with coal mining activities in the nineteenth century.

It can be argued that Pittsburgh's most spectacular set of city steps occupied this area. The Indian Trail steps, long ago consigned to history, consisted of a mile-long set of wooden steps draped over the side of Mount Washington from present-day Carson Street near the Fort Pitt tunnels to a point near the intersection of Shaler Street and Grandview Avenue in the Duquesne Heights neighborhood. While it's no longer possible to follow that historic trail, many photographs survive.

TOURS IN THE WEST END
"West End Village and Elliott: History and Scenic Views"
"Duquesne Heights: Captivating Points of View"
"Off-Road Adventures"

Mere words and numbers cannot do justice to the city steps. Made of materials vulnerable to weather and wear from frequent pedestrian traffic, the stairs are an expression of the past frozen in time, and their imperfections can arouse a sense of tranquility and empathy. Their unconventional beauty must be seen and traveled on to be appreciated. While these guided walks cannot encompass all of Pittsburgh's rich history and contributions, it is our hope that they fuel interest and fascination that prompts additional explorations. Lace up your shoes, and let's go.

PITTSBURGH'S HILLSIDE PLANTS

Matt Holbein is cofounder of the Foliage Library, an artist book series published with Emily Brooks that promotes botanical education through art, geometry, engineering and environmentalism. A Western Pennsylvania native, Matt lives in a desirable location: near the Yard Way city steps in the South Side Slopes. The hillsides of the Slopes offer an ideal spot to view the many native and non-native invasive plants you may encounter as you travel around the city.

"Pittsburgh was once a very deforested place," he explains, "but in the 1930s, the replenishment of trees began, and by the 1950s, many had been intentionally planted. But, as people moved out of the city, plantings became less organized, and native and non-native invasives began to take over."

When asked about the different invasive plants near the city steps, he immediately points to Japanese knotweed. "Knotweed is almost always seen near city steps, and in the height of summer, it can completely overtake a flight and remove it from view." Introduced in the United States as an ornamental plant in the late 1800s, it grows and spreads extremely fast—so fast that most people would never guess it

In warmer months, Japanese knotweed overtakes the Hampshire Avenue city steps in Beechview.

originated outside of Western Pennsylvania. Perhaps the only redeeming quality of knotweed is its clusters of small white blooms. Loved by honeybees, the nectar from these flowers results in a dark, rich honey with hints of maple and vanilla, a unique and tasty find often spotted at regional farmers' markets in the fall. The tender shoots of young knotweed are a bit of a culinary delight reminiscent of rhubarb, but caution is necessary when foraging as the plant absorbs any toxic materials found in the soil or nearby water sources.

"One of the most destructive native invasives," Matt reveals, "is wild grape. It's found throughout Pennsylvania and can take over trees and suffocate them." The vines can reach more than fifty feet because they grow with trees and spread across the canopy. The grapes it produces are less sweet than varieties found in stores, but they are edible if you're lucky enough to find some before the birds do.

The porcelain berry can often be found growing on top of Japanese knotweed and trees. Matt describes it as "similar to wild grapes but easily identified in the fall by its robin-egg blue berries. They're quite beautiful and eye-catching," he says. While these berries are edible, they have a bland flavor.

English ivy frequently covers the hillsides, but it can also climb up to eighty feet and smother trees. "This is one vine that you frequently spy growing underneath flights of city steps," Matt explains, "and it climbs up and over the concrete and railings." Not only can English ivy compromise the structural integrity of the flight, but it can also create a walking hazard.

Many walkers appreciate the delicious scent of honeysuckle, and bush honeysuckle is frequently found near the stairs. According to Matt, "This one is easily identified in the spring and summer by its creamy colored flowers and sweet floral aroma, and in the fall by its bright red berries." Because of these qualities, it may be the most appealing of the non-native invasives.

As with all invasives, a concerted long-term effort is necessary to keep them at bay. "Pittsburgh's hills make it hard to clear invasives, so native plants and trees have the time and space to develop deep root systems," Matt explains. "But trees with deep roots help prevent landslides and erosion, so it's necessary." In some areas, individual residents or neighborhood associations have adopted empty lots and hillsides to undertake remediation and are planting varieties of pine trees, mountain laurel, dogwood and native hibiscus. "It's a lot of work to reclaim the land once invasives take hold," he concludes, "but there are places where it's being done, and the change is noticeable."

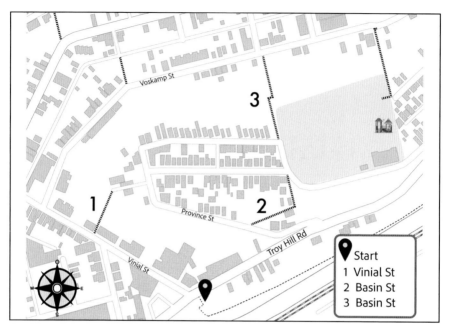

Start
1 Vinial St
2 Basin St
3 Basin St

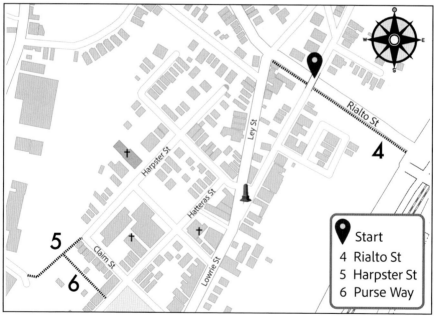

Start
4 Rialto St
5 Harpster St
6 Purse Way

TROY HILL

WHERE EVERYTHING OLD IS NEW AGAIN

Troy Hill is a relatively flat neighborhood that sits on a plateau high above the Allegheny River. With a small, rectangular-shaped footprint less than two miles long by half a mile wide, it also includes Herrs Island within its boundaries. In earlier times, the neighborhood also included the flat river plain that led to the Allegheny River. For much of the twentieth century, this plain, now occupied by the Pennsylvania Route 28 expressway, was a Croatian neighborhood settled by immigrants. Similar to its neighbor Spring Garden, the earliest immigrants who settled here worked in the tanneries, breweries and railroads that lined the Allegheny River.

While Troy Hill has experienced a significant population decline since the 1940s, the neighborhood has recently been making a comeback by revitalizing its parks and greenspaces and attention to small businesses along Lowrie Street. Primarily a quiet, residential area, Troy Hill's close-knit community and easy access to public transportation, bike lanes and city steps appeal to younger, new transplants to Pittsburgh. This walk shows you two distinct areas of the neighborhood. Each walk can be done separately or combined for the whole experience of old and new.

BOUNDARIES: Vinial Street to the west, Voskamp Street to the north, the Lowrie Street bridge to the east and the St. Nicholas Church Pedestrian/Bike Trail to the south.

DISTANCE: Two separate loops, each less than one mile. Combined, they are about three miles.

DIFFICULTY: Easier. Longer stretches of flat walking with occasional flights.

PARKING: Street parking in the neighborhood.

PUBLIC TRANSIT: The 4 bus will take you through Troy Hill with stops at Vinial and Lowrie.

TROY HILL: WESTERN LOOP

Start at the Welcome to Troy Hill mosaic sign at the corner of Troy Hill Road and Vinial Street. International artist and Pittsburgh native James Simon created this glass and ceramic tile artwork. While many of Pittsburgh's ninety neighborhoods have welcome signs at major entrance points, Troy Hill's is among the largest and most vibrant. Next, cross the street to Penn Brewery, initially started as Eberhardt and Ober Brewery in 1883 by William Eberhardt and John P. Ober. William Eberhardt was born in Alsace, France, on April 20, 1844, and immigrated with his parents to Allegheny City (now known as Pittsburgh's North Side) in 1846. Conrad Eberhardt, William's father, was a brewer in Wurtemberg, Germany, and established a brewery in his own name in Allegheny City in 1848.

John P. Ober was born on August 21, 1848, in Allegheny City and started working at his father's brewery at age fifteen, where he was employed for seven years until he joined Eberhardt. Together, they ran the William Eberhardt Brewery until 1883, when they purchased the J.N. Straub Company Brewery and incorporated its holdings as the Eberhardt and Ober Brewing Company.

In 1889, the Eberhardt and Ober Brewing Company allied its interests with those of the Pittsburgh Brewing Company, which it consolidated into in 1899. The Eberhardt and Ober Brewery buildings on Pittsburgh's North Side are listed in the National Register of Historic Sites and have been occupied by the Pennsylvania Brewing Company "Penn Brewery" since 1986.

On your right are the Vinial Street stairs (1), half a block past Penn Brewery. Neighborhood residents and local firefighters frequently use this long, straight flight (177 steps) for training purposes. Regardless of your pace, when you reach the top, look out and over the treetops. In the winter, you'll have a clear view of Mount Washington, downtown, the South Side Slopes and the Lower Hill District. Looking down at the flight, you may be tempted to think of Alfred Hitchcock's classic film *Vertigo* (starring Western Pennsylvania's very own Jimmy Stewart). Follow the sidewalk out to Province Street and look to the left. The First Bohemian Presbyterian Church and parsonage were built in 1908 by the Czech residents of this westernmost section of the neighborhood, which was referred to as "Bohemian Hill." When finished, turn right to walk downhill. After you pass a series of brightly painted murals on the raised foundations of two hilltop houses, turn left to climb the first section of the Basin Street stairs (93 steps) (2). In recent years, this flight and the hillside around it have been under local residents' care. They've actively stewarded the area by building retaining walls, incorporating native plants into the exposed hillside to thwart erosion and keeping the paint looking nice on the stair railings. However, gravity is forever at work, and some sections of the concrete slabs resemble the floors of a carnival funhouse.

Climb halfway to arrive at an overlook with a spectacular view of the valley below and downtown. From this elevation, a viewer can easily see modern skyscrapers and their one-hundred-plus-year-old counterparts. In addition, you'll also see the iconic Heinz 57 stacks. While Heinz's North Side operations ended in 2002, the H.J. Heinz Company had been an integral part of the Troy Hill neighborhood since 1890. Named after its founder, Henry J. Heinz, one of this company's first products was Heinz Tomato Ketchup. The company pioneered processes for sanitary food preparation and led a successful lobbying effort in favor of the Pure Food and Drug Act in 1906. It's not hard to imagine neighborhood residents traveling up and down these very stairs to get to work. Today, Heinz Lofts, a residential complex, occupies six of the former production buildings.

When you reach the top of the stairs, Cowley-Goettmann Park is on your right. With a community center, playground, spray park, basketball court and baseball field, this park may have the best view in the city. Maps from 1862 indicate that an oil refinery once occupied this space and was likely part of the "oil rush," which began in 1859 in nearby Titusville, Pennsylvania. With the park to your right, you'll quickly reach the second segment of the Basin Street stairs (188 steps) (3). As you meander down this green and somewhat overgrown hillside, keep in mind that according to the city maps from 1923, homes once lined either side of the stairs. If your walk

The Basin Street city steps connect Spring Garden to the section of Troy Hill once known as Bohemian Hill.

happens to be during the cooler months when the greenery has died back, you'll spot the remains of stone foundations and the walkways that once led to them. Voskamp Street in Spring Garden greets you at the bottom (see the "Spring Garden and Spring Hill–City View" walking tour to explore this area). Turning left takes you to Vinial in a few blocks, and another left takes you back to Penn Brewery.

Troy Hill: Eastern Loop

Start on the Lowrie Street bridge that looks over Rialto Street and the city steps (4). Rialto Street is the fifth-steepest street in Pittsburgh, with a 25 percent grade. The bridge and city steps below were recently redesigned, so the bridge connects directly to the sidewalk steps below. This vantage point offers a great view of the Troy Hill Incline Company mural created by Phil Seth. While there are only two inclines operational today, the Duquesne and Monongahela in Mount Washington, there were at least twenty others scattered throughout the city during its industrial heyday. These inclines were used for hauling coal, freight and millworkers who had a bit of extra cash and were too weary to trudge up the flights of steps to their homes.

As you head west on Lowrie (mural to your right), turn right on Froman Street. Take a moment to admire the old Troy Hill Firehouse, which the Pittsburgh City Council recently awarded a historic building designation. The building's original terra-cotta plaque spells out in decorative typeface, "Engine Co. No. 11, Erected A.D. 1901," and harkens back to a time of horse-drawn fire trucks and firefighters sliding down poles. Across the street from the firehouse is another Troy Hill mural. *Everyone Belongs in Troy Hill* was created by local artist Brian Gonnella, and it's not uncommon to see people photographing this eye-catching, cartoon-like map of the neighborhood. Continue down Froman and turn left on Harpster Street. Within a block, St. Anthony's Chapel will be on the right. Built in 1880, St. Anthony's holds the unique distinction of housing nearly five thousand Roman Catholic relics, the largest collection of its kind outside the Vatican. Reportedly included are twenty-two splinters from the cross on which Jesus was crucified, a scrap from the Virgin Mary's veil and bones from all twelve of Jesus's apostles. It's a spot that still attracts worldwide pilgrims and curiosity seekers.

At the end of Harpster Street is a flight of city steps (5). You can take the flight (145 steps) all the way down to Lager Street in Spring Garden but turn

Despite the overgrown hillside, the Harpster Street city steps connect Spring Garden to Troy Hill. An old apple tree awaits travelers at the bottom.

left onto Purse Way (48 steps) (6) and note how this is an intersection of two flights of city steps. Harpster and Purse Way have examples of "orphan houses," homes without direct street access that can only be approached by way of city steps. Whereas the house on Harpster now has a driveway that connects it to Lager, the house on Purse Way does not. From the top of Purse Way, turn right onto Hatteras Street. When you reach the intersection of Hatteras and Lager Streets, you'll see an American Legion memorial on a parcel of land. According to the 1923 city maps, a flight of stairs once occupied this parcel that connected the Spring Garden and Troy Hill sides of Lager Street through the hillside. Continue down Lager, and to the left is the Troy Hill Citizens' Community Park, a public space reclaimed from where the Troy Hill Public School once stood. Turn left onto Lowrie Street to return to the Lowrie Street bridge and Rialto Street in just a few blocks.

Troy Hill: Connecting East and West Loops

By looking at any online street map, there are many ways to connect these two walks. To follow the route with fewer vehicles and more of a neighborhood experience, use Lowrie Street and Cowley-Goettmann Park as the connector. If you started with the second tour, instead of turning left on Lowrie to return to the Lowrie Street bridge, turn right and walk several blocks toward Cowley-Goettmann Park. If you started with the first tour, when you reach the top of the Basin Street stairs, enter the park, follow the walkway along the fence and exit onto Lowrie.

It's also possible to connect the two routes using the East Ohio Street bike and pedestrian path. Although a safe and convenient connector worthy of exploration, because of its proximity to Route 28, it's not a quiet walk by any means. This paved trail features interpretive displays of the businesses and community that once occupied the land, including St. Nicholas Croatian Catholic Church. The church was established in 1894 and is the site of the first Croatian Catholic parish in the United States. Unfortunately, it was demolished in 2013 to make room for the expansion of Route 28. The trail is accessed across the street from Penn Brewery (there is a small parking lot for trail visitors) and continues to the Rialto Street city steps (167 steps) (4). Visiting Rialto Street is interesting not only for its legendary steepness but also for its history. Throughout the nineteenth and early twentieth centuries, pigs were unloaded from North Side rail lines and driven up Rialto and down Wicklines Lane to the slaughterhouses in Spring Garden, earning it the nickname "Pig Hill." The Rialto Street city steps were rebuilt in 2022 and feature a built-in bike runnel for easy transport up and down the flight.

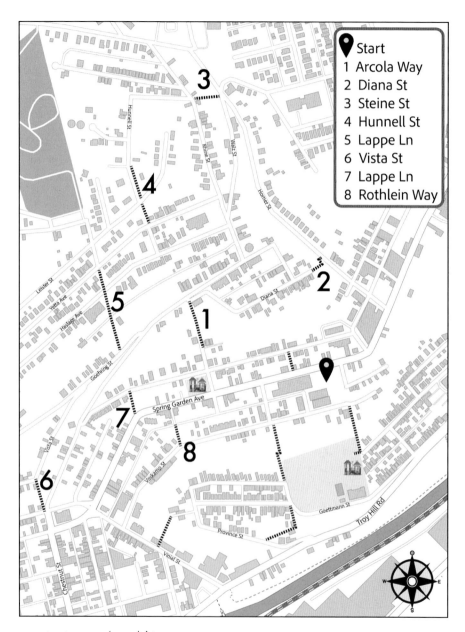

Start
1 Arcola Way
2 Diana St
3 Steine St
4 Hunnell St
5 Lappe Ln
6 Vista St
7 Lappe Ln
8 Rothlein Way

openstreetmap.org/copyright

SPRING GARDEN AND SPRING HILL–CITY VIEW

PUT SOME SPRING IN YOUR STEP

Spring Garden is one of the North Side's oldest neighborhoods. It sits in the crescent-shaped valley between the hilltop neighborhoods of Troy Hill and Spring Hill–City View. Located in proximity to the Allegheny River, the area was once a bustling hub of breweries, stockyards, slaughterhouses and tanneries. Continuing into the present day, this community has long had a mixture of manufacturing and residential uses. Many of the neighborhood's wood-frame homes were built during the Civil War era and housed those who worked in the local industries.

Spring Hill, also known as Spring Hill–City View, is adjacent and north of Spring Garden and was named for its natural underground springs, great views and dramatic hills. Much like others surrounding it, the area was home to German-speaking immigrants. This heritage can be seen in the names of the neighborhood's streets and those buried at St. John's Lutheran Cemetery. As the population swelled in the area, public stairways were constructed within each neighborhood. While many of the original stairways built in the early twentieth century were made of wood, the ones we see today were constructed in the years immediately following World War II. As the summit of Spring Hill–City View has an elevation of 1,066 feet, there are many panoramic views of downtown Pittsburgh and the North Side, and the city steps make for convenient (albeit high-intensity) pedestrian paths for traversing the landscape.

BOUNDARIES: Spring Garden Avenue to the south, Homer Street to the east, Steine Street to the north and Chestnut Street to the east.

DISTANCE: Approximately 3.5 miles.

DIFFICULTY: Moderate to strenuous. The first half is all uphill and involves multiple flights of city stairs.

PARKING: Street parking on any side street. No parking meters.

PUBLIC TRANSIT: The 6 or the 7 bus services the Spring Garden neighborhood.

Start at Threadbare Cider and Mead, located at 1291 Spring Garden Avenue. This building dates from the Civil War era and was formerly owned and used by the Lappe family as a tannery. While the building has been updated to house a modern-day ciderhouse, much of the original brick structure is still in place.

Spring Garden's main road, Spring Garden Avenue, connects to Reserve Township. Because of its location in a valley, this area experienced significant flooding, most notably the Butcher's Run Flood of July 1874, which resulted in 150 deaths.

From the Threadbare Cider parking lot, take a left on Spring Garden Avenue and turn right onto Arcola Way. To the left is the Catalano playground and World War I and World War II Memorial. Straight ahead, there's a long, seemingly never-ending flight of stairs (1). But don't worry, it has "only" 200 individual steps, so it's not even close to being one of the longest in the city. Once you reach the top, turn right on Diana Street and continue along a quiet street of houses. At the end of Diana Street, descend the Diana Street stairs (108 steps) (2). Once you get to the bottom, be sure to cross to the other side of Homer Street to take in the full view of the stairway's spectacular cantilevered construction against the steeply sloping hillside—it's impressive.

Next, walk up Homer Street. The Steine Street stairs are on the left (3), but first, take a moment to cross the street and enjoy the community flower garden and the mosaic mural by local artist Linda Wallen. Depicting Spring Hill's history and German ancestry, it was created collaboratively with local schoolchildren. Looking closely, you can see that city steps are represented in the mural. A short distance beyond, up Damas Street, lies the Voegtly Spring, a natural water source that gave Spring Hill its name

CITY STEPS AS CREATIVE MUSE

The city steps have long inspired creative expression, whether because of the views, the steep hillsides or the various people climbing up and down. The number of artists who have photographed the city steps throughout their history is cause for another book. Some of the most intriguing photographs detailing the evolution of the city's municipal infrastructure reside in the City Photographer Collection. Archived at the University of Pittsburgh and available online, the portfolio features images taken by the Department of Public Works, Division of Photography, between 1901 and 1979. The photos in the collection show Pittsburgh parks, recreation facilities, public facilities, commercial and municipal building exteriors and interiors, mayoral events, traffic situations and general street scenes. It's a fascinating look at the many structures and assets created by the city and the people who built them.

While many artists have been compelled to paint the city's fiery industrial landscapes and its narrow and crowded streets, two artists have spent nearly their entire lives doing so: Ron Donoughe and Cynthia Cooley. Ron Donoughe's residential landscape paintings seamlessly blend the present with the past and capture the light, colors and ambiance of Pittsburgh's neighborhoods. One could just as easily imagine a 1940s steelworker traversing his city steps or quiet streets as one could imagine a young professional with a coffee in one hand and a smartphone in the other.

While Donoughe's work resonates deeply with today's Pittsburgh, this desire to capture the ebb and flow of neighborhood life has a long tradition. For nearly fifty years, nationally acclaimed artist Cynthia Cooley has been known for her vivid paintings of Pittsburgh's hillside neighborhoods and the city steps. Her depiction of the Middle Street stairs in the North Side neighborhood of East Allegheny graces the cover of Bob Regan's *Pittsburgh Steps*, and visitors to the spot often marvel at how her painting highlights all the best features of the landscape. Cooley's work can be found in public, private and corporate collections worldwide.

Linda Wallen's ceramic mural on the Vista Street city steps connecting East Allegheny and Spring Garden.

There's also an assortment of artists whose creative work has appeared on the city steps themselves. In 2019, the local cycling and pedestrian advocacy organization BikePGH collaborated with Pittsburgh's Office for Public Art to match local artists with neighborhood organizations in West End Village, Fineview, Polish Hill and Troy Hill to beautify and draw attention to their city steps. While the city's policies for permanently modifying its public stairways require several levels of approval, two neighborhoods, the South Side Slopes and East Allegheny, successfully incorporated mosaic murals into stairways. A mural by Linda Wallen is featured on the Vista Street stairway, and Laura Jean McLaughlin's work is featured on Oakley Way. Both locations are eye-catching and popular spots for photographs.

Filmmakers and multimedia artists have also found inspiration in the winding hillsides and their stairways. Notable short films shown locally and online in recent years include Catherine Drabkin and Pahl Hluchan's *Steps in Motion: A North Side Animation*, highlighting the Arch Street stairs in the Central Northside; *Pinburgh*, a short animated fantasy by Doug Cooper, Andrew Mellon Professor of Architecture at Carnegie Mellon University; *To Heaven and Back: Troy Hill*, an audio-visual documentary walk by Erin Anderson and Danny Bracken through the neighborhood; *The Best Urban Hiking Is in Pittsburgh?*, a film by Dean Bog that features dozens of flights of city steps; and Paola Corso's *On the Way Up: City Steps, City Immigrants*, featuring immigrants and immigrant descendants living in Pittsburgh. The filmmaking process inspired Corso to write *Poems and Photographs of City Steps* (Six Gallery Press, 2020), which describe the working-class lives of the many immigrants who traveled the stairs every day.

It's also important to note that while many of the individuals referenced above may consider themselves full-time artists, an ever-growing number of people share their city steps–inspired artwork, photography, poetry and writing on social media. This creative participation by visitors and residents of all ages, backgrounds and interests shows the commanding presence the city steps have for all who travel their concrete and steel surfaces.

Above: Steine Street (*left*) and Diana Street (*right*) city steps in Spring Hill–City View.

Opposite, top: The original Diana Street steps were constructed from wood in 1908. *Pittsburgh City Photographer Collection, University of Pittsburgh.*

Opposite, bottom: The Voegtly Spring, a natural water source that gave Spring Hill its name more than one hundred years ago.

more than one hundred years ago. As a quick side excursion, a short walk farther up Damas Street brings travelers to Castle Damas—an enormous retaining wall constructed in 1933 to look like a medieval fortress.

As you approach the Steine Street stairs, take a moment to look around. Your view today is not very different from what was captured in a photograph of the area in 1911. While the old Engine Company No. 53 firehouse is now a boxing gym, there's a historical photo of this intersection of Homer and Walz Streets from the early 1900s that looks very similar to today (except for the dirt road and lack of trees). In fact, the photo shows wooden stairs in the same spot where the concrete ones of today stand.

Once you climb the Steine Street stairs (105 steps), continue walking up a steep section of Steine Street. Before reaching the very top, turn onto Hunnell Street for one last surge in elevation before reaching Erk

Above: The view of downtown Pittsburgh from the top of the Hunnell Street city steps (shown in the bottom left).

Opposite: The cantilevered Lappe Lane city steps strike an impressive pose in the section that connects Goehring Street and Zang Way.

The Lappe Lane city steps have several sections, with this flight starting at Spring Garden Avenue.

The Vista Street city steps viewed from the end of Chestnut Street in East Allegheny (also known as Deutschtown).

Way/Tank Street. From your lofty position, take a moment to look out over the horizon. Perspectives like these give the neighborhood its name: Spring Hill–City View.

As you walk down Hunnell Street, a wooden flight of sidewalk stairs (55 steps) (4) takes you down to Leister Street. Turn right and then turn left on Lappe Lane. The city steps begin at the intersection of Yetta Avenue (5). Traveling down Lappe Lane, cross Haslage Avenue and Zang Way before arriving at Goehring Way. If you're out walking during the fall season, keep your eyes open for "monkey balls." About the size of a baseball but with a bumpy, yellow/green exterior, these are fruits of the Osage orange tree. While the fruit is inedible for humans, monkey balls were a favorite snack of the woolly mammoths that traversed the area thousands of years ago.

Once you reach Goehring Street, cross the street and look up to the Lappe Lane stairs you just descended (222 steps). Like the flight at Diana Street earlier in this walk, this section of Lappe Lane is also cantilevered against the hillside and creates a striking presence. Once you're done admiring the construction, continue walking uphill along Goehring and, at the fork in the road, head left to descend along Vista Street. The road becomes very narrow, with old houses built into the hillside, and the Vista Street city steps are nearby on your left (6). This flight was completely rebuilt in 2021, replacing a flight that dated back to the late 1940s. The previous flight had fallen into significant disrepair, and the city closed it. However, based on the neighborhood's popularity and proximity to public transportation, the flight was selected for a complete redesign. As you walk down this flight, notice the many modern construction practices not employed on flights from eighty years ago: a bike runnel track that allows cyclists to walk their bike along the stairs, a lower profile for each step, enclosed railings, crumble-guards on the lip of each step and LED lighting for nighttime safety. When you reach the bottom, you're greeted by another Linda Wallen community mosaic mural. This one depicts historical scenes from the Spring Garden and East Allegheny neighborhoods.

If you look down Chestnut Street toward downtown, you'll see a portion of the street that's paved in red brick. Until 2024, the street retained embedded trolley track lines, a throwback to a time when trolleys traveled along many of the Spring Garden and Spring Hill–City View streets. The Spring Garden Line opened in 1915 and served the community until it was closed in 1957, the same year the Spring Hill Line stopped service.

From Itin Street, turn left onto Spring Garden Avenue. As you travel the several blocks back to Threadbare, to your left and right are additional flights of stairs. Another section of Lappe Lane appears on the left (7) and Roethlein Way (8) on the right. This area is rich in steep hillsides, so you don't need to travel far to find more city steps to explore.

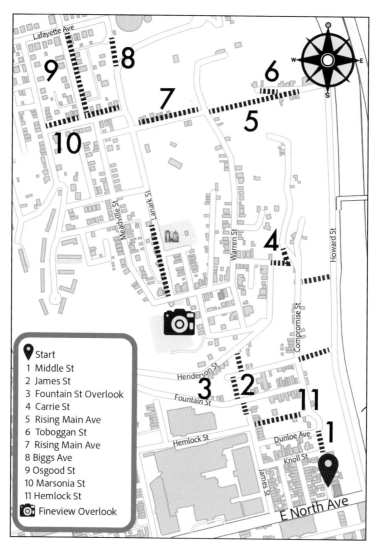

Lafayette Ave

9

8

7

5

6

10

Meadville St

Lanark St

Warren St

Howard St

4

Compromise St

Start
1 Middle St
2 James St
3 Fountain St Overlook
4 Carrie St
5 Rising Main Ave
6 Toboggan St
7 Rising Main Ave
8 Biggs Ave
9 Osgood St
10 Marsonia St
11 Hemlock St
Fineview Overlook

Henderson St

3

2

11

Fountain St

Hemlock St

Dunloe Ave

Knoll St

1

James St

E North Ave

FINEVIEW

LIVING UP TO ITS NAME

Perched behind Allegheny General Hospital on Pittsburgh's North Side, Fineview is home to some of the best vistas in town. It's hard to believe, but back in 1788, surveyor David Redick visited the area, and his report to then president of the Commonwealth's Supreme Executive Council Benjamin Franklin was unfavorable: "[It] abounds with high Hill and deep Hollows, almost inaccessible to a Surveyor.…I cannot think that ten-acre lots on such pits and hills will possibly meet with purchasers, unless like a pig in a poke it be kept out of view."

As much as we disagree with Redick today, his original assessment rang true at the time. The land was a nearly impassable series of hills separated from the main settlement of Allegheny by lowlands and swamps, with no "fine view" to behold. Nevertheless, in 1828, the nuns of the Order of St. Clare erected the St. Clare Young Ladies Academy on a sixty-acre site in the area, but it wasn't until the second half of the nineteenth century that residential development began. For this reason, the neighborhood was initially known as Nunnery Hill and featured the Nunnery Hill Incline (1888–95), which connected Federal Street to what is now Meadville Street. However, by 1915, the nuns and convent were long gone, prompting the local board of trade to lobby the Pittsburgh City Council to replace the neighborhood's old, irrelevant name in favor of one that might boost local real estate and business activity: Fineview.

Today, Fineview lives up to its name as a lovely hillside neighborhood with a rich history and spectacular views of downtown Pittsburgh. There are houses that have survived since the Civil War era, community gardens, forested greenways and recreational parks. But most importantly, Fineview is known for its many sets of city steps, most notably Rising Main, which has 337 steps.

BOUNDARIES: Lafayette Street to the north, Henderson Street to the east, Meadville Street to the west and Knoll Street to the south.

DISTANCE: Approximately 3.25 miles.

DIFFICULTY: Above average. Hilly terrain, periods of sustained uphill climbing combined with stretches of flat walking. Adding on the optional Rising Main walk is more challenging.

PARKING: Street parking throughout the neighborhood.

PUBLIC TRANSIT: The 11 or 8 bus will take you to Fineview.

This walk will take you along sections of the Fineview Fitness Trail, the Fineview Neighborhood Trail and the Fineview Step Challenge. For online maps, see the "Resources" section at the end of the book.

Start at Middle and Knoll Street, where the Fineview Fitness Trail begins (1). As you ascend the Middle Street steps (101 steps), look behind you at the fantastic view of the Central Northside neighborhood and downtown. Whatever the season, this view never gets old.

Turn left on Dunloe Avenue and then right on James Street, go past the parking garage for Allegheny General Hospital and ascend the James Street stairs (62 steps) (2), which also provide a great view from the top. Allegheny General Hospital began as a fifty-bed infirmary in 1886, and while various additions and enhancements occurred in subsequent decades, in 1920, the New York architecture firm York and Sawyer designed one of the nation's first "skyscraper" hospitals, which you can still see today. The cornerstone for the building was laid in 1930, but the Great Depression interrupted construction. The twenty-two-story, $8 million hospital eventually opened its doors in 1936.

Turn left on Fountain Street and then up the next set of James Street stairs (89 steps). You'll reach Graib Street, and to the left, you'll see the entrance to the Fountain Street Overlook. This 2.2-acre hillside park has a short, wooded trail that offers views of Mount Washington at one end and the

The Middle Street city steps mark the start of the Fineview Fitness Trail.

downtown skyline at the other. As you meander along, you'll see the remains of stone foundations, evidence of the eighteen houses that once clung to this steep hillside and were torn down in the 1950s and '60s. The Overlook has a flight of concrete stairs connecting to Fountain Street (3). For a shorter walk, you can descend, take a left on Fountain and retrace your journey. To continue, return to Graib Street and immediately climb the stairs on your left (56 steps).

At the top, turn right on Henderson Street and continue your uphill climb. At the intersection of Jay Street, you'll see the historic Henderson-Metz House, built in circa 1860 by Robert Henderson and featuring a Gothic Revival style. The property commands an all-encompassing view of downtown Pittsburgh and the Allegheny River Valley. Continue up Henderson until you reach the intersection with Carrie Street. To your right are the Carrie Street stairs (106 steps) (4). These lead down to Sprain Street and are worth exploring for the flight's unique combination of stairs and raised catwalks to traverse the hillside. If we were walking this area along Sprain and Compromise Streets in the early twentieth century, several additional flights of stairs and paper streets would have connected us to hillside roads. Today, those options are no longer available or are in a treacherous condition, so reverse your direction and exit out onto Carrie Street and continue past the community garden. When you reach Warren Street, take a right and continue for a few blocks through this residential neighborhood featuring a mix of new, old and older homes, many of which have front porches but few of which have driveways or garages. Being a pedestrian can be a challenge on old, narrow streets such as these, as you'll frequently find vehicles parked on sidewalks. Exercise caution and do the best you can.

When you reach the intersection with Rising Main Avenue and look to the right, you'll see a long flight of stairs. You're looking at the "legendary" Rising Main (5), the fourth-longest flight in Pittsburgh (331 steps) and the second-longest flight still intact and walkable from top to bottom. The current condition of the flight is stable but significantly affected by hillside erosion exacerbated by the demolition of abandoned properties that once occupied the area and underground water seepage caused by decaying water and sewer pipes. Like the Sprain Street stairs, this is another "down and back" excursion that can be explored independently or as part of the entire tour. If you choose to descend, you'll quickly note the "spinal scoliosis"–like quality owed to erosion and time. As you look over the valley, you'll see Highway 279 and another steep hillside that lies beyond. This area was once a neighborhood known as East Street Valley. Flights such

The Carrie Street city steps wind down the hillside to Sprain Street. The West End Bridge is visible in the distance.

as Rising Main were constructed in the early twentieth century to lead hilltop residents to the industry that lined the valley. However, starting in the 1960s and continuing through the 1980s, highway construction to connect outlying suburbs to urban areas increased, and nearly one thousand families and two hundred businesses were closed or relocated as they were within the proposed rights-of-way. As you reach the bottom of the flight, you'll come to Howard Street, a now-quiet road parallel to the Thruway. At the base of

Left: A section of the Howard Street city steps in the East Street Valley (1934). *Pittsburgh City Photographer Collection, University of Pittsburgh.*

Below: The Royal Street city steps look down from Brahm Street into the East Street Valley neighborhood (1932). *Pittsburgh City Photographer Collection, University of Pittsburgh.*

Opposite: Several sections of elevated sidewalk steps are located along Marsonia Street.

Rising Main Avenue and Toboggan Street (which also has a raised flight of 78 sidewalk stairs) (6) is a reservoir pumping station. The first station opened in 1882, and while the present building is not the original, it has maintained its original function. It's hard to imagine what the area would have looked like a century ago, but historical photos such as those from Historic Pittsburgh, an online archives compilation, offer a glimpse into the past.

As you return to the top of Rising Main, continue up Rising Main Avenue and feel free to walk the 31 wooden sidewalk steps (7) that line part of the road. Turn right onto Biggs Avenue and continue along to the intersection of Glenrose Street, where, on the left, you'll see a Fineview Fitness Trail sign and a flight of stairs (53 steps) (8) snaking through the hillside. Once you reach the top, the reservoir will be to your right. At the top of the hill, turn left on Lafayette Street and then left to descend the Belgian block–paved Osgood Street. You'll notice a raised sidewalk on the left side that gradually rises from the street below as you walk down the block (9). Fortunately, you'll begin to descend as you reach the intersection with Marsonia Street. You'll reach street level again after descending a short flight of stairs.

It's worth crossing to the other side of Marsonia to see the arrangement of raised sidewalks and city steps (10). As you turn left on Marsonia, you'll notice that the retaining walls of the sidewalk could use some care and that

This page: The Biggs Avenue city steps in 1928 compared to the same area today. Because of Pittsburgh's dense urban forest, this view has changed considerably.

Opposite: The Lanark Street sidewalk steps lead to the Fineview Overlook with its "fine view" of downtown.

Another impressive downtown vista, this one courtesy of the Hemlock Street city steps.

the condition of the walkway tends to fluctuate, but it's still fascinating. When you reach Biggs, turn right and, passing Rising Main Avenue, continue to Lanark Street. You'll pass a telecommunications tower to the left. In the Nunnery Hill days, the land now enclosed in fence and wire once held St. Mary's Cemetery and Mortuary Chapel. The cemetery was relocated to Mount Troy Road in Reserve Township a few years later, and the chapel was demolished in 1913.

On your left is the Fineview Playground, which features a covered pavilion, a half basketball court, a running track and a variety of playground equipment pieces for kids of all ages. Most notable is the inclusion of a pickleball court. For those unfamiliar with the sport, it's a hybrid of tennis, badminton and ping pong played with a paddle and plastic ball. As you progress along Lanark, you'll once again come to a raised sidewalk, but the real attention grabber is directly in front. As you reach the end of Lanark, cross Catoma Street and enter the Fineview Overlook. The Overlook is a perfect spot to take a breather and see a spectacular view of Pittsburgh. As you marvel at the landscape and think back to the hilly terrain you've traveled, you should not be surprised to learn that in an earlier day, Fineview had a local streetcar line.

As you exit, turn right on Catoma Street, and at the intersection of Myler Street, you'll see Heathside Cottage. Stonemason and engineer James Andrews bought some of the first residential land in Nunnery Hill in 1864 (perhaps he knew it had "Fineview" potential) and commissioned the building of the Gothic Revival house. It was listed in the National Register of Historic Places in 1974 and is currently a private residence that offers occasional public events.

As you continue down Catoma, you'll soon reach the intersection with Warren Street. For those of a certain age who are familiar with the 1983 film *Flashdance*, this is the intersection featured in the title sequence where Alex rides her bike down the hairpin turns of the hillside. Turn right onto Warren, and after two blocks, turn left on Jay Street. At the intersection with Henderson, turn right and go back down the James Street steps until you reach the Allegheny General Hospital Parking Garage. Looking to the left, you'll see the Hemlock Street stairs (143 steps) (11). This quiet and shady flight borders houses to the left and overlooks the city on the right. You'll ultimately exit onto Compromise Street, where you'll turn right onto another hairpin curve for Dunloe Avenue. Round the corner, and the Middle Street stairs, along with their fine view, will be right in front of you.

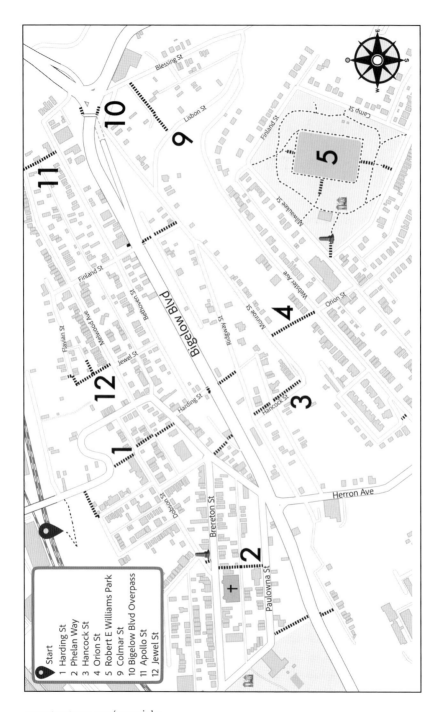

Start
1 Harding St
2 Phelan Way
3 Hancock St
4 Orion St
5 Robert E Williams Park
9 Colmar St
10 Bigelow Blvd Overpass
11 Apollo St
12 Jewel St

Blessing St
Lisbon St
Finland St
Camp St
Milwaukee St
Webster Ave
Orion St
Finland St
Bethoven St
Mallwood Ave
Flavian St
Jewel St
Harding St
Ridgway St
Monroe St
Hancock St
Bigelow Blvd
Dobson St
Brereton St
Paulowna St
Herron Ave

openstreetmap.org/copyright

POLISH HILL AND THE
HILL DISTRICT'S UPPER HILL

A 360-DEGREE PITTSBURGH EXPERIENCE

Polish Hill

Today, Polish Hill is an enclave of hillside houses and budding businesses nestled along a nearly vertical quarter-mile expanse just east of downtown Pittsburgh and rising above the Strip District. But in the early nineteenth century, the area was called Springfield Farms and owned by a handful of prosperous landowners. As the area along the Allegheny River became more urbanized, German and Irish immigrants settled, with the first Polish immigrants arriving in the late 1800s.

By 1905, when the iconic Immaculate Heart of Mary Church was built, the neighborhood had become known as Herron Hill. While many residents were Polish, Irish and German, African American families also lived in the area. Most residents belonged to the working class, laboring in the steel mills or rail yards at the bottom of the hill.

In the late 1960s, residents petitioned the city to call their neighborhood Polish Hill, and the name has stuck ever since. As you walk around this urban neighborhood with the feel of a close-knit village, you'll encounter families of Polish ancestry who have lived here for generations, as well as newer residents attracted to the neighborhood's small size, walkability and proximity to jobs and public transportation.

Polish Hill's Brereton Street sidewalk steps, with the Immaculate Heart of Mary Church in the distance.

UPPER HILL

Upper Hill is the farthest east portion of the Hill District and borders Polish Hill and North Oakland. Like Polish Hill, Upper Hill has undergone several transformations in Pittsburgh's history. In the 1870s and 1880s, the neighborhood centered on Herron Avenue was called Minersville. Little of this neighborhood survives except for the old cemetery near Milwaukee Street. While the Upper Hill Reservoir was constructed in 1880, the surrounding area was not immediately developed for residential housing, despite the area's pastoral feel and amazing views. In time, this section of the neighborhood became known as "Sugar Top," a place where countless Black professionals—barbers, teachers, lawyers, ministers, doctors and owners of jazz clubs and restaurants—settled. The Upper Hill even experienced a touch of national fame. Lena Horne briefly lived here, and Thurgood Marshall was known to stay in the neighborhood when he visited the city.

Today, Upper Hill is a mix of homeowners and renters attracted to the proximity of hospitals and universities in North Oakland and the incredible views available on the north- and east-facing sides of the neighborhood.

The Harding Street city steps are viewed from Herron Avenue near the East Busway transit stop.

BOUNDARIES: MLK East Busway to the north, Finland Street to the east, Bigelow Boulevard to the south and Herron Avenue to the west.

DISTANCE: Approximately 4.0 miles + optional loops.

DIFFICULTY: Moderate. The first half is uphill and involves multiple flights of city stairs.

PARKING: Street parking in Polish Hill and Upper Hill. Meter and time-restricted parking along Bigelow Boulevard.

PUBLIC TRANSIT: Multiple public transit options exist through the East Busway, Bigelow Boulevard and Centre Avenue.

This walk begins on Herron Avenue at the Herron Station of the East Busway and can be completed as a loop or as a one-way walk ending in North Oakland. There are also two side excursions that can be incorporated or walked separately. There's so much to see and explore here, and walkers have many options available.

From the exit of the East Busway, make your way up Herron Avenue. There will be a flight of stairs to your right, but for this walk, you'll want to take the set directly in front of you. At 80 steps, this section of Harding Street (1) is an example of a mapped, official street that is only accessed via steps. In fact, there are approximately 344 streets like this in Pittsburgh. Once you reach the top, you'll be on Dobson Street in Polish Hill. Take a right. Like many other older neighborhoods, Polish Hill once had thriving, vibrant sidewalk-level businesses. While much of that has disappeared with time, this block continues to attract independent shops appealing to residents and visitors.

As you continue down Dobson to the intersection at Brereton Street, you won't be able to miss the Immaculate Heart of Mary Church. As one of the oldest and largest churches in the city, it has quite a commanding presence and may be one of the most photographed churches in Pittsburgh. Sandwiched between the Polish Hill Civic Association and the church's rectory (the building adjacent to the church which houses clergy), you'll see a short flight of stairs (7 steps) (2) leading to a brick walkway. Walking this path leads you to Phelan Way, directly at the back of the church.

• • • • • • • • • • • •

Outsider Art on the Hillsides

Since 2015, the *Pittsburgh Orbit* blog has explored the often-unseen corners of Pittsburgh's art, history and culture. Launched by local musician, writer and website developer Willard Simmons, the *Orbit* has shared photographs and stories about ghost houses (an imprint of a no-longer-standing house on one that is still here) and "ghost signs" (the long-gone practice of painted wall advertisements), strange acts of black and gold sports fandom, street art, curiosities and interesting facts about regional cemeteries and the city steps. Simmons is undoubtedly "nebby" (see the Nebby N'At on page 199) when it comes to city steps. Over the years, he has traveled the stairs, written about unique aspects of their environs and pondered the proliferation of street art on empty hillsides near the stairs and on existing infrastructure.

"Over the years, I've found many interesting intentionally placed metal sculptures, graffiti, tin-can paintings and other creations," he says. "Some things appear to be one-offs, while others are a series, located throughout a neighborhood or even the entire city."

One of Simmons's favorite places is the hillside along the Harding Street stairs between Bigelow Boulevard and Ridgeway Street in Polish Hill. He describes the sculptures and paintings in the area as giving "the impression that someone, or a small group of people, decided they wanted to decorate the place." Despite repeated attempts over the years to find the artists behind the work, their identity remains a mystery.

Some neighborhoods display public murals, monuments or sculptures condoned or commissioned by the City of Pittsburgh. Simmons mentions artist Linda Wallen, for example, who installed some "beautiful community-generated mosaic murals" in Spring Hill–City View on the Vista Street stairs and the Damas Street parklet. But in this vibrant urban environment, artistic expression also knows no bounds. "Lots of Linda's ceramic racing pigeons, a nod to the hobby of the area's German settlers, can be found all over the

Above: Examples of anonymous DIY art are located alongside the Harding Street city steps.

Opposite: The trellis at the Carrie Street city steps in Fineview.

neighborhood," he says. "They're on retaining walls, on the city steps, Jersey barriers—all over the place."

For Simmons, the open spaces in Pittsburgh's depopulated areas have fostered a rich DIY art scene. Infrequently used or abandoned places became a canvas for local creativity, out of public view and unlikely to be taken down or defaced. "These pieces of art have a small audience," he explains, "but they tend to stick around for a long while and ultimately become part of the landscape."

One of his favorite flights of city steps is the Carrie Street stairs in Fineview, where a city-approved trellis sculpture welcomes you at the entrance on Henderson Street. "That stairway and hillside feel so magical," Simmons says. "You enter by walking under the trellis, and then the flight has all these turns and catwalks as you make your way down to Sprain Street. The hillside is beautiful, and the stairs are an engineering feat. It's one of my favorite places to take people because it feels like you've entered a place that is not in the regular world."

Simmons cherishes his years walking the city steps and documenting curiosities found along the way. He has an

appreciation for Pittsburgh's unique infrastructure and all the history behind it, with an understanding that "the city steps will likely never be built in such a widespread way ever again. There's melancholy and magic in knowing that."

SIDE TRIP OPTION (.15 MILES)

Continue down Brereton Street for a short distance until you arrive at 30th Street. The West Penn Community Rec Center, playground, spray park and skate park will be in front of you. The current West Penn Recreation Center was built in 1929–30, but before that, the area was the site of West Penn Hospital, Pittsburgh's first chartered public hospital. Founded in 1848 and opened in 1853, the four-story, 120-patient hospital also held the Western Pennsylvania Medical College and Training School for Nurses. The city of Pittsburgh served as a transportation hub during the Civil War, and sick and wounded U.S. troops were treated at West Penn Hospital from 1861 to 1865.

By the turn of the twentieth century, the hospital was outdated, and a new hospital was constructed in the Bloomfield neighborhood. The land was sold to the city, and it wasn't long before the Community Recreation Center and grounds were built. The former hospital and school were located on the present site of the ball field and West Penn Park.

Across the street from the West Penn parking entrance is Phelan Way. Walk a short distance along the back of the church until reaching the brick walkway on the left.

• • • • • • • • • • • • •

Once you're on Phelan Way, take a right to continue up the next flight of stairs (13 steps) to the long-empty Immaculate Heart of Mary school. The empty brick walls are a popular spot for local and visiting graffiti artists and contain some stunning artwork. This is also a popular spot for "urban exploration" photography (also known as urbex). While the city steps will take you up to Paulowna Street (60 steps), for this walk, trace your footsteps back to Phelan Way, turn right and walk the length of this narrow street uphill until you reach Herron Avenue. Turn right, and at the intersection, cross to the other side of Herron across Bigelow Boulevard.

As you cross the street, you may not realize that a pedestrian underpass once connected both sides. After it sustained structural damage due to a vehicle crash in 2023, PennDOT closed and filled the tunnel with concrete. Throughout its many decades of use, walking the tunnel was rarely a pleasant experience due to litter, foul smells

The Phelan Way city steps with the Immaculate Heart of Mary Church in the background. Graffiti fills the walls of the now-empty school.

and poor water drainage. However, the walls were always colorful with graffiti and worth braving the noxious conditions to see during the drier months.

Continue up Herron Avenue and turn left on Ridgeway Street. Ridgeway is the border of Polish Hill and Upper Hill—running right down the center of the street. Ridgeway still has several old brick rowhouses, and it won't be long before you've transitioned from the housing density of Polish Hill to the more pastoral vibe of Upper Hill.

Turn right on Hancock Street and take the sidewalk stairs (66 steps) (3). Of the city's 900-plus flights of public stairs, approximately 280 are sidewalks. These stairs help navigate the elevation change and are safer than walking in the street. When you reach the top, be sure to turn around and look out over the view. Below Polish Hill, you'll see the Allegheny River and, across from that, the North Side neighborhood of Troy Hill and the 31st Street and 40th Street bridges.

The view from the top of the Orion Street steps includes the Allegheny River and points north.

A winter view of the Sunday Street sidewalk stairs in California-Kirkbride. Union Dale Cemetery is on the hilltop.

One of the longest flights in Pittsburgh, the Yard Way city steps in the South Side Slopes have an abundance of natural beauty.

The Vinial Street city steps in Troy Hill offer a wintertime view of downtown.

Fall foliage surrounds the Diulus Way city steps in Central Oakland.

The illuminated city steps of South 18th Street connect the South Side Flats to the South Side Slopes.

Garvey's Bar anchors the bottom corner of the Sterling Street city steps in the South Side Slopes.

Milroy Street, and its city steps, marks the boundary between Perry North and South neighborhoods.

The view from the St. Michael Street city steps in the South Side Slopes. An "orphan house" appears on the left.

Mount Washington's Cuthbert Street city steps offer a view of modern downtown skyscrapers and old neighborhood churches.

The Frazier Street city steps in Greenfield connect neighborhood residents to public transportation on Greenfield Avenue.

Despite being crooked with age, parts of the wooden stairs on Hunnell Street in Spring Hill–City View are still functional.

The Colmar Street city steps viewed from Blessing Street in the Upper Hill neighborhood.

The Leolyn Street city steps overlook Carrick. The St. Basil Church is visible on the distant hilltop.

While Rising Main Avenue is one of the longest flights of city steps, it also offers picturesque views from its wooden sidewalk stairs.

The Sumner Street city steps in the South Side Slopes connect to Barry and Holt Streets' sidewalk steps. The downtown skyline is in the distance.

The Kerr Street city steps in the West End. The old stone foundations of homes become visible during the fall and winter months.

The former St. Michael's Church and the Hartford Street city steps in the South Side Slopes. Wooden stairs lead to the Knoxville Incline Greenway.

The view of Pittsburgh's North Shore and beyond from the top of the Elbon Street city steps in the West End.

One of the longest flights of city steps in Pittsburgh, a view of Fineview's Rising Main from Overbeck Street in Spring Hill.

The Oakley Way city steps in the South Side Slopes feature a mosaic mural by local artist Laura Jean McLaughlin.

If you think this view is fantastic, just wait until you see what's waiting for you at the very top of Upper Hill.

Turn left onto Monroe Street, and then to the right will be the Orion Street city steps (126 steps) (4). Climb these to the top, cross Webster Avenue and continue to Milwaukee Street. At the corner of Milwaukee is Madison Elementary School. Built in 1902, the school was expanded in 1929 to serve the neighborhood's growing population. Walking up the steps of the school from the street puts you close to the ornately decorated doors and windows featuring real and fantastical creatures. The school was listed in the National Register of Historic Places in 1986 and is currently being developed as a performing arts space for the Pittsburgh Playwrights Theatre.

Turn left on Milwaukee Street, and the Robert E. Williams Memorial Park and Reservoir is on the right. When it comes to Pittsburgh's reservoir parks, Highland Park Reservoir in the Highland Park neighborhood is the oldest, largest and most popular with visitors. In contrast, Williams Park, named after Pittsburgh's first African American police detective and magistrate, is low-key and secluded. There's a small, relatively modern playground and a well-preserved World War I memorial created by Frank Vittor (who also created the Honus Wagner statue at PNC Park). This plot of land, one of the highest points in Pittsburgh, has held a reservoir since 1872. It is considered a secondary reservoir, drawing its water from the Highland Park Reservoir. There are several flights of stairs in the park, some made of stone slabs and others of concrete (5). It's worth climbing to the paved track circling the reservoir, as it's here that you'll see and experience a near-360-degree view of the Pittsburgh landscape. The views are farthest reaching in the colder months when the foliage has dropped, but there's always plenty to experience. Walking around the park's street-level perimeter, you'll notice red brick streets that speak to a time before asphalt paving was standard. While streets like these are now preserved for historical significance, residents like them for their natural "speed calming" tendencies.

For those who want to continue to the Upper Hill section of Schenley Farms, the next side trip option will lead you there. If you're not going to Schenley Farms, take a little R&R at Williams Park and read ahead for the return trip to Polish Hill in the following paragraphs.

• • • • • • • • • • • •

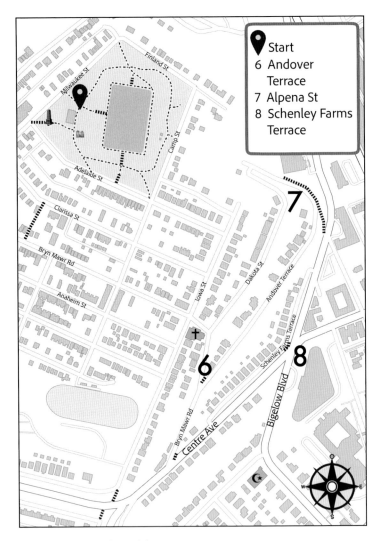

Start
6 Andover Terrace
7 Alpena St
8 Schenley Farms Terrace

openstreetmap.org/copyright

SIDE TRIP OPTION

From Milwaukee Street, turn right on Finland Street and then right onto Camp Street. Walk along the tree-lined residential street until you reach Adelaide Street. Turn left and continue to the end when you turn right onto Iowa Street. At the intersection is the historic Grace Memorial Presbyterian Church. Grace Memorial is the oldest African American Presbyterian church in Pittsburgh and

The double set of stone city steps leads to Schenley Farms Terrace from Centre Avenue in Upper Hill.

Allegheny County. It was founded by Henry Highland Garnet, a former enslaved person who migrated to Pittsburgh in 1868 as the new president of Avery College, an institution of higher learning for African Americans located in Allegheny City (now known as the North Side). As a child, Garnet and his entire family dramatically escaped enslavement in Maryland. As an adult, Garnet devoted himself to ministry and support of abolitionist causes. While Grace Memorial had previous locations, the church has remained at this location since 1948.

Turn down Bryn Mawr Road and marvel at the old Belgian block pavers. While you might be tempted to refer to the street's pavers as cobblestones, do refrain—cobblestones are generally round river stones, while Belgian block pavers have been cut and shaped into rectangles. The difference is subtle, except for the price (cut block pavers would have cost more).

While much of the historic neighborhood of Schenley Farms is in North Oakland, a slice exists in the Hill District bordered by Bryn Mawr Road and Andover Terrace. As you turn left onto Andover, walk down a small corner flight of four steps (6). Yes, this is an official flight of city steps. There are dozens of flights with five steps or fewer scattered throughout the city's streets, parks and playgrounds.

The view of the railroad tracks and valley below the Jewel Street city stairs.

As you walk down this narrow street of attractive large homes, you'll catch the occasional view of the rest of the neighborhood farther down the hill.

While the University of Pittsburgh now owns a section of Schenley Farms, there are still more than one hundred homes that date back to the late 1890s and early 1900s. As you reach the end of Andover, you'll see a flight of stairs to the right (7). Travel down these forty-seven steps to Bigelow Boulevard. Turn right, and in a short time, you'll be at the entrance to Schenley Farms Terrace, a narrow, red brick–paved street filled with historic homes (and the cars that belong to them often parked on the sidewalks!). About halfway down this street, you'll come to two sets of stairs that connect to the intersection of Bigelow Boulevard and Centre Avenue (8). This flight is best appreciated from the bottom, where it's easier to see the wide stone slabs that form each tread. This flight appears on the city's historical maps from 1923, so it stands to reason that they were built once development in the neighborhood warranted them. Continue to the end of Schenley Farms Terrace, and you'll turn right onto Centre Avenue. Continue up Centre until you reach the Belgian block pavers of Bryn Mawr (and the tiny flight of four steps on the corner). Then, start your slow ascent of retracing your travels back to Williams Park.

• • • • • • • • • • • •

THE RETURN TRIP TO POLISH HILL: After all that uphill walking, this downhill trip is a special treat. From Williams Park, retrace your steps past the Madison school and turn right on Webster Avenue. As the street begins to S-curve, turn left on Lisbon Street, a narrow passage etched into the steep hillside. To the right are the Colmar Street stairs (120 steps) (9). When you reach the bottom, turn left on Blessing Street and approach Bigelow Boulevard. Travel up the pedestrian overpass (10) and, upon descending and exiting, turn left on Bethoven Street. If you're curious about the spelling of this street, you're not alone. In 1991, local classical music station WQED-FM petitioned to have the street renamed Beethoven to match the spelling of the composer's name, assuming that it had been misspelled. Residents complained against the change, and the matter was dropped. However, recently digitized street maps from 1923 indicate the original spelling to be Beethoven, so the radio station's claim had merit. Bethoven Street curves again to the left, and after passing an auto body shop, the Apollo Street stairs (98 steps) (11) are on the right. Descend to Melwood Avenue and turn left. As you pass the many homes with vibrant and eclectic front porches, after 0.25 miles, you'll come to a flight of city steps on your right. The Jewel Street stairs (12) lead to two small streets running parallel to Melwood, and it's worth exploring these 163 steps if only for the views of the hillside and Allegheny River. The freight rail line runs along the bottom of the hill. You'll know a train is coming by the sounds that fill and reverberate against the hillsides.

Once you're done exploring the Jewel Street stairs, continue down Melwood and turn right on Herron. You'll soon be back at the entrance to the Busway (or, if you decide to visit one of Polish Hill's legendary watering holes for a bit of refreshment, there are a few to choose from).

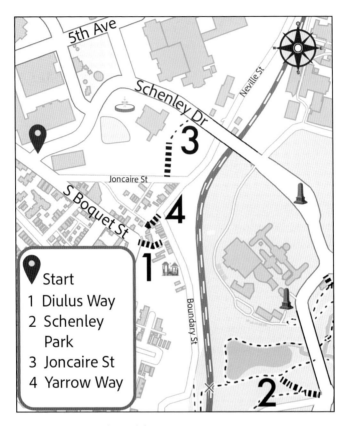

5th Ave

Schenley Dr

Neville St

3

Joncaire St

S Boquet St

4

1

Boundary St

2

Start
1 Diulus Way
2 Schenley Park
3 Joncaire St
4 Yarrow Way

OAKLAND

PANTHER HOLLOW MAGIC FOR MUGGLES AND MARAUDERS

Oakland is Pittsburgh's academic, cultural and healthcare hub. Divided into four separate neighborhoods—north, south, central and west—it's home to three different universities, plus museums and parks, hospitals and a seemingly unlimited number of shops, bars and restaurants. Oakland got its name from the oak trees found on the farm of William Eichbaum, who first settled in the area in 1840, and Oakland Township developed rapidly following Pittsburgh's Great Fire of 1845. By 1860, there was considerable development along Fifth Avenue, and in 1868, Oakland was annexed to the city of Pittsburgh.

Central Oakland, where most of this walk takes place, contains the enclave of Panther Hollow. Italian immigrants settled in this small, somewhat isolated neighborhood at the bottom of Joncaire Street in the late nineteenth century. Even today, park benches and picnic tables are painted in the Italian flag's red, white and green, and a remembrance memorial commemorates the original ninety-five families who first settled here. The Pittsburgh Junction Railroad first laid tracks through this area in the 1880s, and a century later, the rails became the Junction Hollow Trail, part of Pittsburgh's extensive biking/hiking network.

Panther Hollow, which is named for the mountain lions once native to the area, also shares a border with Schenley Park, Pittsburgh's first intentionally designated park. Mary Schenley's gifts of acreage and land purchases made by the City of Pittsburgh helped to establish Schenley Park, technically considered part of neighboring Squirrel Hill, in 1889. In time, this area would also include Schenley Plaza, which features industrialist Andrew Carnegie's library, art and natural history museums, as well as a concert hall complex. Schenley Park is a designated historic district in the National Register of Historic Places.

BOUNDARIES: Schenley Drive to the north, South Bouquet Street to the west, Juno Street to the south and Panther Hollow Road to the east.

DISTANCE: Approximately 2.0 miles.

DIFFICULTY: Light to moderate. The first half is all downhill, then a gradual uphill grade through Schenley Park. Much of this walk has sidewalks.

PARKING: Street parking can be challenging in Oakland. However, metered parking is available on select side streets.

PUBLIC TRANSIT: The P3 and the 54 bus service Fifth Avenue, close to South Bouquet.

Start on South Bouquet Street at the intersection of Roberto & Vera Clemente Drive. For those unfamiliar with Pittsburgh's rich professional baseball history, Clemente spent eighteen seasons playing for the Pittsburgh Pirates, with thirteen as an All-Star. Clemente was outstanding both on and off the field and spent the off-seasons providing equipment and food to Latin American and Caribbean countries. Sadly, he died in a plane crash in 1972 while delivering aid to earthquake victims in Nicaragua. A museum dedicated to Clemente's baseball career, personal life and humanitarian causes is located nearby in the East End's Lawrenceville neighborhood.

Turn left on South Bouquet Street and enter Central Oakland, a residential area that many University of Pittsburgh (locally referred to as "Pitt") students call home. The buildings in this area are a mix of old and new, but of note is 379 South Bouquet, once the Saint Lorenzo di Gamberale Mutual Benefit Association clubhouse. The association was established in 1917 to provide mutual aid to families in need, and the clubhouse was constructed in 1938 and served into the 1990s. Today, the building is an apartment complex.

A few doors down from the former clubhouse, you'll find the entrance to the Diulus Way city steps (99 steps) (1). As the architecture and interior design of Pitt's Cathedral of Learning are often compared to Harry

Central Oakland's Diulus Way meanders down into Panther Hollow.

Potter's Hogwarts, the labyrinthine quality of Diulus Way is reminiscent of Diagon Alley. This enchanting entrance sets the stage for exploring the Panther Hollow section of Schenley Park and its stone slab public stairs. After journeying down Diulus and exiting right onto a quiet area of Boundary Street, you'll see one of the entrances to the Junction Hollow Bike Trail to the left.

FOR THOSE LOOKING TO ADD TO THIS WALK, THIS JUNCTURE PRESENTS TWO OPTIONS. The first is to continue along Juno Street (a dirt road to the right) to see the underside of the Anderson Bridge (see "Off-Road Adventures"). The second is to continue straight on the Junction Hollow Trail, a flat, 0.6-mile one-way hike that connects walkers and bikers to the other side of Boundary Street in Four Mile Run and the Schenley Park soccer field.

To continue without these add-ons and access Schenley Park and Panther Hollow Lake, turn left at the Junction Hollow Trail and look for a path on the right that leads across the railroad tracks. Carefully cross the tracks, and the lake will be directly ahead. Initially constructed in 1892 as part of the early development of Schenley Park, Panther Hollow Lake was designed for rowing and skating. It was originally ten to twelve feet deep, but the lake's high sedimentation rate has reduced its current depth to two feet. The Pittsburgh Parks Conservancy is working to reduce sedimentation and improve water quality to support aquatic life and migratory birds.

Follow the path around the water's edge counterclockwise until you see the old stone slab city steps (2). For more than a century, park-goers have traveled up and down these flights, which are downright magical. Rustic and rugged, they snake along the wooded hillsides in complete harmony with their surroundings.

The late Victorian–era romantic landscape of the park is primarily credited to E.M. Bigelow, Pittsburgh's first director of public works, and landscape architect William Falconer, who served as park superintendent until 1903. The rugged terrain of Pittsburgh, combined with their creative vision, produced dramatic, winding roads, stunning open amphitheaters and broad vistas. Rugged-looking man-made features—such as stone retaining walls and staircases and stone or wood bridges, fences, pavilions and shelters—were also included.

Near the top of the hillside, the wide stone slab stairs transition to modern concrete ones and exit onto Panther Hollow Road. At the top of the stairs, turn left and approach the Panther Hollow Bridge. Originally constructed in 1895–96, it features four bronze sculptures by Giuseppe Moretti of panthers crouching as sentinels on each bridge corner. After crossing the bridge and marveling at the views, turn left onto Schenley Drive and feast on the visual smorgasbord of the Phipps Conservatory and Botanical Gardens.

After passing the expansive grounds of Phipps, cross over the Schenley Bridge. The bridge was completed in 1897 as part of the main entrance to Schenley Park, and its design is similar to the nearby Panther Hollow Bridge. However, whereas that bridge features Moretti's panther sculptures,

The view of Panther Hollow Lake from the stone steps leading to Panther Hollow Road.

this bridge is known for its large number of love locks placed on the wire security fence. While other bridges in Pittsburgh contain these symbols of love, which typically include the sweethearts' names or initials and date, this spot has the most. In addition, the Schenley Bridge features a clear view of

The Yarrow Way sidewalk steps on a short but steep area of Panther Hollow.

the boiler plant referred to as the Cloud Factory in Michael Chabon's 1988 debut novel, *The Mysteries of Pittsburgh*.

At the end of the bridge, take the first left onto the road that says "University of Pittsburgh Frick Fine Arts." For an optional side trip, continue along Schenley Drive. To the right is the main branch of the Carnegie Public Library, and to the left is the Mary Schenley Memorial Fountain. Also known as *A Song to Nature*, this 1918 landmark sculpture in bronze and granite was designed by Victor David Brenner (best known for the Lincoln penny). Across the street, Schenley Plaza has numerous public amenities and attractions.

To return to Panther Hollow and walk the Joncaire Street city steps (139 steps) (3), walk to the rear of the Frick Fine Arts Library building. This flight, originally constructed of wood in the early 1920s and rebuilt in concrete in 1949, connects pedestrians and cyclists from the highly traveled streets near Schenley Plaza to Panther Hollow below.

The flight you see at Joncaire Street today was completely rebuilt in 2018. Unlike the wooden flights from the turn of the twentieth century (version 1.0) and the concrete stairs constructed to replace them after World War II (version 2.0), the 3.0 iteration of city steps is very different. Most notably, they're broader and deeper but with a shorter rise, making the climb a bit easier on our twenty-first-century bodies. In addition, the railings are constricted, creating extra safety precautions for children and dogs. The lighting is better, and the hillside landscaping is designed to cope with water runoff. Finally, and most interestingly, a runnel track is placed to the side for cyclists to get from the nearby Junction Hollow bike trail to all points throughout Oakland and Squirrel Hill. These are fantastic improvements, even if you still feel the burn walking up or down.

When you reach the bottom, the Belgian block–paved Joncaire Street awaits. Turn left and then right onto Boundary Street to make a return trip up "Diagon Alley" (Diulus Way), or turn right and explore the old and narrow ways of Isis and Yarrow. Yarrow Way (19 steps) (4) reconnects with Joncaire farther up the hill and intersects at South Bouquet Street near Roberto & Vera Clemente Drive.

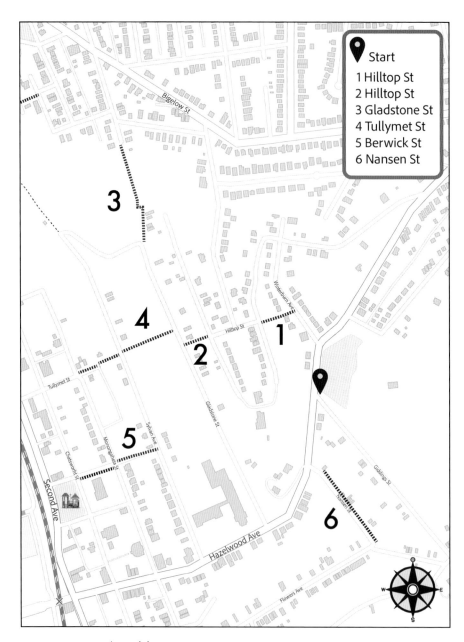

Start
1 Hilltop St
2 Hilltop St
3 Gladstone St
4 Tullymet St
5 Berwick St
6 Nansen St

HAZELWOOD

WITH EVERY STEP, ANOTHER STORY UNFOLDS

Legend has it that Hazelwood got its name by combining the hazel of the hazelnut trees that were once plentiful in the region with the surname of the area's founder, Colonel George Woods. We don't know for certain if that's true, but it makes for a good anecdote. In fact, Hazelwood has no shortage of interesting stories, and this walk explores several of them.

Throughout its long history, several historical notables have crossed paths with Hazelwood, but perhaps the most underappreciated is Hugh J. Ward. His name may not ring a bell, but the game he created nearly a century ago, bingo, is still played by all age groups at parties, in social halls and in community centers around the country. Ward modeled his version of the legendary game of chance after a Canadian game called "Housey-Housey." Liking the game's concept but disliking the name, he rebranded it using the then popular expression "Bingo" for marketing and advertising purposes. The game quickly caught on and has enjoyed widespread, lasting appeal—especially for senior residents who attend the Hazelwood Active Living Center most Friday afternoons.

With its current population hovering around 4,900, Hazelwood's green and rugged hillsides and winding country-esque roads make a visitor feel as if they've arrived in a small rural town. Up until the 1890s, Hazelwood had only limited connections to the larger city environment of Pittsburgh and was considered a pastoral enclave for the wealthy. However, from 1870 to 1950, the neighborhood's population skyrocketed from 1,399 to 13,000 without any change in its geographical boundaries. Jones & Laughlin (J&L

Steel), the Baltimore & Ohio Railroad (B&O) and many other businesses involved with iron and steel production or fabrication precipitated this job growth. The area transformed into a place far from idyllic, as heavy industry brought smoke and fumes that hung in the air and negatively affected the environment. In recent years, remediation efforts have taken place in Hazelwood and in other riverfront neighborhoods to undo decades of destructive pollution, while former industrial sites are transforming into green technology and energy hubs.

BOUNDARIES: Winterburn Avenue to the east, Gladstone Street to the north, Chatsworth Avenue to the west and Flowers Avenue to the south.

DISTANCE: Approximately 2.5 miles.

DIFFICULTY: Light to moderate. The walk fluctuates between uphill and downhill with occasional flat stretches. Much of this walk has sidewalks.

PARKING: Street parking is plentiful. Be mindful of driveways and signage for any restrictions.

PUBLIC TRANSIT: Many options, including the 93 and the 56 bus, operate throughout the heart of Hazelwood. While Hazelwood Green and Mill 19 are not included in this walk, they are within walking distance and can be accessed by the 57 bus.

This walk begins at the intersection of Giddings Street and Hazelwood Avenue near Gladstone Field. Take a right on Hazelwood Avenue, and after the field, turn left on Winterburn Avenue. Many of the homes in this section date back to the 1940s and '50s and are designed with a low and broad rectangular profile, a central chimney and a pitched, side-gabled roof. The Hilltop Street city steps (84 steps) appear on the left (1). This paper street appears on the 1923 historic street maps (see the sidebars "The Ways of Pittsburgh" and "Finding Gaps in the Railings" to learn more) but has only ever contained city steps. Once you reach the top, continue along Hilltop and then descend to Parnell Street. A short wooden flight leading to a path is right in front of you (2). The path leads to another flight of wooden stairs (27 steps) that lead down to Gladstone Street. If walking in the summer, there may be some initial overgrowth—a good sign that you're headed in the right direction.

Once you reach Gladstone Street, take a right, and soon city steps appear on your left. For a brief side excursion, continue down Gladstone Street until you reach the end and then climb the Gladstone Street city steps (140 steps) (3). This flight connects Hazelwood with Greenfield and travels through part of the Hazelwood Greenway. Houses on the right side of the stairs date to the early 1900s. As you walk up the stairs and along the catwalks, try to imagine a time when homes densely lined both sides of the stairs, which provided residents access to the streets below. Long before this area was a "greenway," it was a fully developed and densely inhabited area. When you reach the end of the stairs and reconnect with Gladstone Street, turn around and retrace your steps.

When you return to the vicinity of Hilltop Street, you'll find the Tullymet Street city steps on your right (4). Descend to Sylvan Avenue, cross the street and then descend a second flight (259 steps) to Chance Way. From here, you'll have a view of the Hazelwood Green Riverfront Trail Park. The riverfront trail is a National Park Service Community Partner of the Lewis and Clark National Historic Trail Experience. Nearby is Mill 19, a business development center focused on advanced technology research in artificial intelligence, automation, robotics and other innovations. On this site, James Laughlin built the Eliza blast furnace along the Monongahela River in 1859. A few years later, he partnered with B.F. Jones, who had rolling mills across the river on the South Side. Together they formed Jones & Laughlin, or J&L Steel Company. In 1952, the company constructed a $17.5 million (more than $181 million today) bar rolling mill on the Hazelwood site. Following the decline of Pittsburgh's steel town heyday, J&L's Hazelwood Works was bought by Ling-Temco-Vought Incorporation (LTV) in 1974. Within two decades, the plants closed and ended operations, signaling the end of industry on the site.

Throughout the early 2000s, a significant amount of environmental remediation and site prep occurred on the site, bringing it to its current state. In August 2020, then Democratic candidate and former vice president Joseph R. Biden held his first major speech at Mill 19 after formally accepting his party's presidential nomination. Pittsburgh native and Golden Globe–winning actor Michael Keaton also filmed an interview for *60 Minutes* here.

As you reach the intersection of Tullymet and Monongahela Streets, one of Pittsburgh's oldest surviving structures, the John Woods House is on the left. Built in 1792, this stone house is believed to have been built by Colonel George Woods, the surveyor who laid out Pittsburgh's Golden Triangle (in present-day downtown) in 1784. Recognized for its historical

This page and opposite: The Gladstone Street city steps in Hazelwood highlight hillside homes and the unique view of the Cathedral of Learning.

Left: The Tullymet Street city steps near the John Woods House. The building was included in the National Register of Historic Places in 1993.

Right: The Berwick Street city steps as seen from Chatsworth Avenue. An additional set connecting to Sylvan Avenue can be seen in the background.

and architectural importance, the Woods House was added to the National Register of Historic Places in 1993. It is now known as the John Woods House and Historic Pub.

Turn left on Monongahela Street and travel for several blocks. As you come to the intersection of Berwick Street, there are sidewalk flights of city steps leading in both directions (5). Take the flight going downhill (54 steps), and soon you'll see the Lewis playground and athletic park at the intersection of Berwick and Chatsworth Avenue. This space was first used as a children's farm in 1910; a decade later, it was transformed into a playground. The playwright August Wilson (1945–2005) spoke of playing basketball here in the late 1950s when his family lived in the neighborhood.

Take a left on Chatsworth Avenue and walk past the park to St. Ann's Roman Catholic Hungarian Church. Constructed between 1919 and 1925, this site was one of three congregations serving the large influx of Hungarian

Hazelwood's Nansen Street city steps are pictured with a Little Free Library and community canned goods cupboard.

immigrants throughout the first half of the twentieth century. The church was closed in 1993 after seventy years.

To the left of the church is the Three Rivers School, an independent school for children ages five to nineteen. The building was originally built in 1954 by architect D. Roderick Jones as the Hazelwood Moose Lodge, and in the 1970s, it served as Hazelwood's YMCA. However, in the early 1900s, this block of Chatsworth Avenue was home to the Thomas Institute, one of the area's early substance abuse treatment facilities for men. Founded by Eli Thomas in 1904 during the height of the temperance movement, the institute offered accommodations for forty patients and was often filled to capacity with a long waiting list.

A colorful flight of stairs leads from the school through the Hazelwood community garden and exits onto Monongahela Street. Take a right and continue down Monongahela toward Hazelwood Avenue. Before you reach the intersection, you'll come to the original Carnegie Public Library of Hazelwood on the right side. Construction began on the building in 1899, and it originally featured a unique dome rising from an octagonal base above the circulation desk. Hazelwood's was the only Pittsburgh library with such a feature. The library also had a music hall and performance space.

THE HAZELWOOD GREENWAY

Pittsburgh's Greenways system was developed in 1980 to systematically deal with abandoned properties under city ownership after the steel industry's collapse. These areas are often found along steep hillsides; many are landslide prone, and some include abandoned mine drainage sites, excessive dumping and informal hunting activities.

Greenways are often large tracts of land, and addressing their environmental issues poses an overwhelming challenge for neighborhood volunteers without funding. Today, the oversight of Greenways is jointly administered by Pittsburgh City Planning and several nonprofit organizations specializing in environmental remediation.

Caring for a greenway is a bit like a large jigsaw puzzle; each piece is unique but connected to others. To prevent ecological collapse, each solution must consider factors such as deer overpopulation, invasive worm species, soil destabilization and landslides caused by increased rainfall.

According to the Pennsylvania Department of Conservation & Natural Resources, Pennsylvania's white-tailed deer population has grown from about ten deer per square mile in the 1700s to an estimated thirty deer per square mile today. Residents and visitors to Pittsburgh often comment on the prevalence of deer in the city, and the empty hillsides near the city steps are where many live and forage for food. However, these white-tailed deer eat young plant shoots before their roots anchor to the ground, contributing to erosion and destabilization of the hillsides.

As deer run rampant, the population of Asian jumping worms expands. Attracted to deer feces and leaf litter, these invasive creatures are destructive to the ecosystem because they rapidly devour organic matter that provides the forest layer critical for seedling and wildflower growth. Without enough leaf litter, soil then washes away during rainstorms.

In 2020, Pittsburgh City Planning received a $50,000 grant from the Trust for Public Land to pilot a new greenway management model for the 183-acre Hazelwood Greenway, the city's most extensive greenway (see map on page 152). In 2021, the pilot expanded thanks to a $430,000 grant from the National Recreation and Park Association. The funding allowed for collaboration between Hazelwood's neighborhood association and environmental groups such as LandForce, Tree Pittsburgh and Allegheny GoatScape to collectively guide the improvement and sustainability of the greenway.

Since 2020, combined efforts have created drainage pathways to mitigate further erosion, targeted invasive vegetation for management and removal, planted native trees to restore and protect the urban forest and constructed a half-mile walking trail. While the Hazelwood Greenway trail is not included in the city steps walk, one entrance is from Longview Street, not far from Gladstone Field on Giddings Street, where the walk begins. Signs mark the entrance to the trail.

August Wilson credited this library as the place that changed his life and gave him proof that it was possible to be a writer. Hazelwood is mentioned in two of his plays, *Joe Turner's Come and Gone* (1984) and *Gem of the Ocean* (2003). The library relocated to newer facilities in 2004, and the building has stood empty ever since. In recent years, advocates have lobbied for its restoration and return to public use.

At the intersection with Hazelwood Avenue, turn left and continue up the street. Within a block, you'll come to the historic Gladstone School. Named to the National Register of Historic Places in 2021, the structure was originally built in 1913 as an elementary school, with a junior high added in 1926 and a high school in 1958. It was the only school in the Pittsburgh Public Schools system to educate students K–12 in a single location, and it was the last school August Wilson attended before he dropped out in 1960. In 2001, population decline led to its closure. The building was recently sold to a community organization with plans to transform it into a housing development.

Walk up Hazelwood Avenue until you reach Nansen Street on the right (6). This street is closed to vehicles, but walkers can use the sidewalk steps (152 steps) to access the area; it's worth exploring for an up-close view of the destruction caused by Pittsburgh's landslides. The city steps along Nansen are also fascinating for their use of raised sidewalks towering several feet above the street.

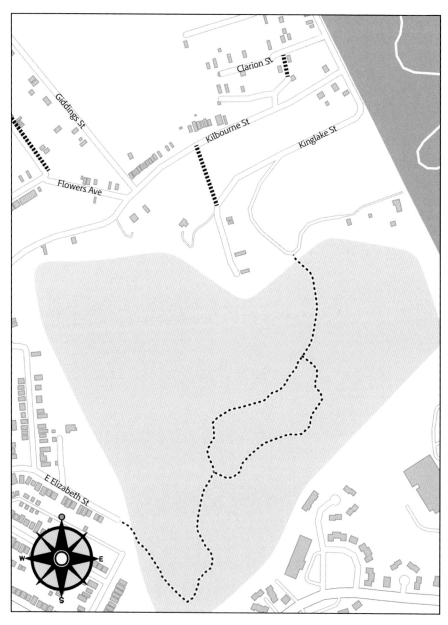

• • • • • • • • • • • • •

SIDE TRIP OPTION

For those who would like a slightly longer walk (by approximately 0.3 miles) with a decidedly rural feel, at the end of Nansen, turn left onto Flowers Avenue. At the end of Flowers, turn left onto Kilbourne Street, followed by another left onto Giddings Street. With only a few houses on this third-of-a-mile stretch, it's easy to forget you're in the middle of a city, and it's also easy to see why Colonel George Woods and the thousands of residents who came after were so attracted to this area.

• • • • • • • • • • • • •

Otherwise, retrace your path up Nansen and turn right on Hazelwood Avenue, and within a block, you'll see Giddings Street and Gladstone Field on the right.

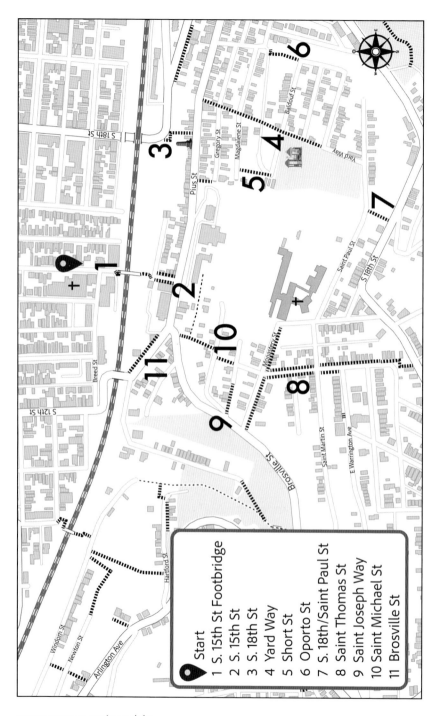

Start
1 S. 15th St Footbridge
2 S. 15th St
3 S. 18th St
4 Yard Way
5 Short St
6 Oporto St
7 S. 18th/Saint Paul St
8 Saint Thomas St
9 Saint Joseph Way
10 Saint Michael St
11 Brosville St

openstreetmap.org/copyright

SOUTH SIDE SLOPES WEST

STAIRWAYS TO HEAVEN

John Ormsby, now known as the "father" or "founder" of Pittsburgh's South Side, arrived in America in 1752 from Dublin, Ireland. After stays in New York and Philadelphia, he traveled west to join General Edward Braddock's army in 1755. Ormsby was involved in the construction of Fort Pitt and served many roles there, including paymaster, until 1762.

In 1763, Ormsby received 2,400 acres along the south bank of the Monongahela River near the present Smithfield Street bridge as payment for his services throughout the French and Indian War (1754–63). In time, this land was divided into four boroughs—South Pittsburgh, Birmingham, East Birmingham and Ormsby—and was ultimately annexed by the City of Pittsburgh in 1872. Today, these areas are known as the South Side Flats and the South Side Slopes. While Ormsby died in 1805, many streets on the South Side still bear the names of Ormsby's children and grandchildren, including Sarah Street, Jane Street, Mary Street, Josephine Street and Sidney Street.

While the South Side Flats is a popular destination for dining, shopping and entertainment, the neighborhood doesn't have a significant elevation change. That is not the case with the South Side Slopes, which has sixty-eight sets of city steps and approximately 2,500 steps (not counting those in parks and playgrounds), more than any other neighborhood in the city.

In the early to mid-nineteenth century, the South Side was one of the largest glass producers in the United States, and glass tableware from the area was used in the White House under Presidents Andrew Jackson and James Monroe. In time, iron and steel mills replaced glass as the primary industry in the South Side and along the Monongahela River. In 1854, B.F. Jones and James Laughlin became business partners and formed American Iron Workers, later known as J&L Steel. By 1929, J&L was producing 1.74 million tons of steel yearly—much of it along the banks of the Monongahela near the Flats.

With this rich industrial history, it's no surprise the Pittsburgh Convention and Tourist Bureau touted South Side industry as the "Workshop of the World." The city steps were used by workers who lived on the Slopes to reach their jobs in the steel mills and glass factories in the Flats. As you walk around, you'll notice that the street names speak to the rich religious traditions of the workforce that immigrated to Pittsburgh during this time. There are also many churches dotting the hillsides, some still used for worship and others repurposed for residential or secular use, their steeples reaching skyward.

ORPHAN HOUSES

Orphan houses are only accessible from city steps. The nearest street may be a few steps or several dozen steps up or down a steep hillside. When these houses were initially built, typically between the late 1800s and 1930s, street access wasn't an issue, as residents often didn't own a car, let alone multiple cars, as we do today. Many orphan houses have since been demolished, and those still standing are likely unoccupied. Some have maintained residents over the decades, but not many. Active orphan homes are often in higher-density neighborhoods, such as the South Side Slopes. This hillside area was once home to steelworkers and their families but is now inhabited mainly by young professionals and college students. The popularity of the Slopes is due in part to its proximity to the entertainment and nightlife of the South Side Flats. The Slopes are also close to the many colleges and universities scattered from downtown, Uptown and Oakland, all easily accessible by one of the many bridges that connect the South Side to the East End or by public

The Behring Street city steps connect to a few remaining "orphan houses" in the South Side Slopes.

transportation. In other neighborhoods—such as Central and Upper Lawrenceville, Stanton Heights or Troy Hill—owners have taken advantage of now-empty hillsides to create access roads or driveways leading directly to their property, but that's the exception rather than the norm.

BOUNDARIES: Breed Street to the north, Yard Way to the east, South 18th Street to the south and South 12th Street to the west.

DISTANCE: Approximately 2.0 miles.

DIFFICULTY: Moderate to challenging. Significant climb in elevation and many flights of stairs. Yard Way has 317 steps.

PARKING: Street and metered parking is available throughout the South Side Flats. Consider parking along South 13th Street near Armstrong Field.

PUBLIC TRANSIT: The 48 and 51 bus lines service East Carson Street in the South Side Flats.

This walk travels along streets and stairs appearing in the South Side Slopes Neighborhood Association's Church Route and Step Trek. For more information, see the "Resources" section at the end of this book.

Start at St. Adalbert's Church (1889) on South 15th Street, an active Roman Catholic church and part of the Mary Queen of Peace Parish. Continue up South 15th to the former Polska Szkola, which sits on the left at the base of the steel footbridge. This school for Polish immigrants was built in 1898. Climb the 59 steps to the footbridge (1). Rebuilt in 2002 at the request of the South Side Slopes Neighborhood Association and residents, this footbridge was completed by the city and Norfolk Southern Railroad to reinstate pedestrian access between the Slopes and Flats. Climb another fifty-nine steps to Clinton Street and then travel twenty-six sidewalk steps to reach St. Michael's Church on Pius Street (2).

The Roman Catholic Church of St. Michael was consecrated in 1848 in a house where the chapel now stands. In 1849, a deadly cholera plague infected many throughout the region. Parishioners prayed to St. Roch, the patron saint of the sick, and vowed to keep a day holy when good health was restored. As the epidemic ended, German parishioners donated land, and Charles Bartberger designed a church in Rhineland Romanesque style, similar to rural Bavarian churches, which he built between 1855 and 1860. While the church has since closed and was redeveloped as the Angels' Arms condominiums, the Feast of St. Roch, also known as Cholera Day, is still observed every August at nearby St. Adalbert's Church.

Turn left on Pius Street, where you'll pass the All Wars Memorial. Dedicated by the Hillside Veterans Association to residents of the South Side who served in the armed forces, the site hosts annual Memorial Day services. For a side excursion, take a walk up and down the South 18th Street stairs (140 steps in each direction) (3). These steps, upgraded and repaired in 2012 through a grant from Duquesne Light, feature LED lights below each tread. Now illuminated, the steps are not only safer for pedestrian travel but also shine like beacons for passersby, highlighting and celebrating this neighborhood icon.

After passing the South 18th Street stairs, Yard Way appears on the right (4). Notice the street sign with the icon of a person walking on stairs—they're not just city steps; they're also an official street. Yard Way is one of the longest flights in the city (see the "Six Longest Flights" list), and with 317 steps, it intersects with seven different streets. Traveling up or down this flight is not for the faint of heart or foot. It's not necessary to climb

The view from the Oporto Street city steps in the Billy Buck Hill section of the South Side Slopes.

all segments at once, though, and the narrow streets of this residential neighborhood are worthy of exploration. Locally referred to as Billy Buck Hill, the name dates to a time when a local store kept Billy goats tethered in its backyard. Today, goats do occasionally appear on the hillsides as part of invasive species remediation efforts, and they are well suited to the rugged terrain.

Sandwiched between Winters Park and South Side Park, the cross streets of Billy Buck Hill connect to several flights of stairs worthy of investigation. Start by climbing Yard Way, and at the second intersection, turn right on Magdalene Street. At the end of the block, the South Baldauf/Short Street (125 steps) (5) stairs are to the left. Climbing these will take you alongside Winters Park and bring you to South Baldauf Street. Travel down this narrow, residential street passing the intersection of Yard Way (don't worry, you'll return to it shortly!) until arriving at Oporto Street (124 steps) (6). Walking down this flight provides an amazing view spanning Oakland, Uptown and parts of downtown.

Once you've reached the bottom of the stairs, walk downhill and around the corner to turn left on Magdalene Street. Within a block, you'll return to Yard Way and be able to continue your ascent, crossing Roscoe, South Baldauf, Huron and Shamokin Streets. Once you've reached the top, be sure to turn around to take in the view of Uptown, Oakland and Greenfield across the river. Continue to St. Paul Street and turn right.

As you travel along St. Paul Street, a narrow flight of city steps on the left leads to South 18th Street (76 steps) (7). However, stay on St. Paul Street, and to the right is the St. Paul of the Cross Monastery of the Passionists Order. Known as the barefoot missionaries, the Passionists vowed to live a life of prayer, poverty, penance and solitude. Also designed by Charles Bartberger, this Romanesque-style monastery was built in 1859 on eleven acres. The grounds may be open for quiet visitation.

St. Paul Street crosses Monastery Street and turns into St. Martin Street. Walk two blocks uphill and turn right on St. Thomas Street. There will be sidewalk stairs on both sides of the street ranging from sixteen to seventy-seven steps long (8). Due to the grade of the hillside, the sidewalk on the left will be of a different construction style than the sidewalk to the right. While it's tempting to focus your attention on the various construction methods employed to create sidewalks and city steps, don't forget to look up and out to the horizon. This section of the neighborhood offers several heart-swelling views of downtown Pittsburgh and The Point.

The last section of St. Thomas Street has a flight of six sidewalk steps. At St. Joseph Way, take a small detour to the left to travel the seventy-seven St. Joseph Way city steps that pass a few "orphan houses" (see the "Orphan Houses" sidebar for a complete description) (9). These homes are not too far removed from the street, and the view is worth the extra walking.

Retrace your steps and continue down the St. Thomas Street–St. Michael Street stairs (193 steps) (10), crossing Hackstown Street. Near the Hackstown entrance, another two stairs-locked homes appear on the left. As you descend the steep hillside, take a moment to revel in the altitude you share with the top of St. Michael's Church steeple. Also visible from this elevation are the Monongahela River and sections of the University of Pittsburgh's campus in Oakland. At the bottom of the stairs, cross the intersection of Pius and Brosville Streets and take a left, walking downhill. The sidewalk stairs on the right side have a railing down the center of the flight (42 steps) (11), a unique feature not found elsewhere in the city. Don't forget to look up and out for heavenly views of the downtown skyline.

Turn right at South 12th Street and walk the overpass above the freight train tracks. Turning right on Breed Street, take note of the massive stone retaining wall to the right. This structure first appeared on the neighborhood street maps of 1910 and still bears the dark soot and ash stains of the industrial air that once blanketed the city. Continue along Breed Street for five blocks to return to South 15th Street and St. Adalbert's Church.

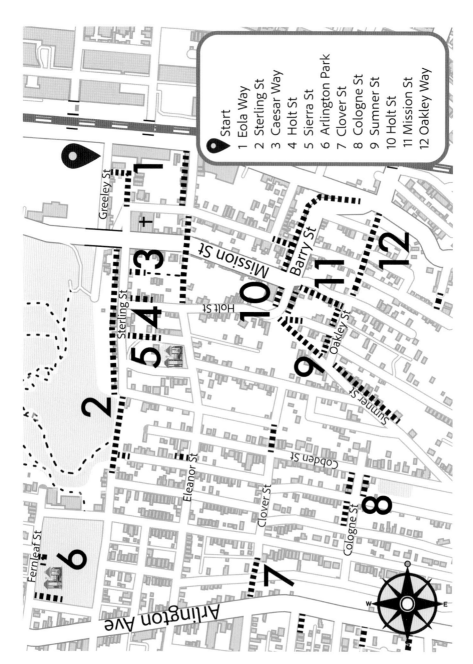

Start
1 Eola Way
2 Sterling St
3 Caesar Way
4 Holt St
5 Sierra St
6 Arlington Park
7 Clover St
8 Cologne St
9 Sumner St
10 Holt St
11 Mission St
12 Oakley Way

Greeley St

Sterling St

Mission St

Barry St

Holt St

Oakley St

Sumner St

Eleanor St

Cobden St

Clover St

Cologne St

Fernleaf St

Arlington Ave

SOUTH SIDE SLOPES EAST

STERLING EFFORTS YIELD REWARDS

This second walk through the South Side Slopes travels through the eastern section of the neighborhood. For the western section, see the "South Side Slopes West: Stairways to Heaven" walking tour.

The streets of the South Side Slopes are legendary for three reasons: their steepness, their narrowness and their city steps. With elevations ranging from 791 to 1,174 feet, several places are too steep for vehicles. In these places, the public stairs are still used to move residents from their homes to work or public transportation. Given this challenging topography, it's hardly surprising that the South Side Slopes has the most city steps in Pittsburgh (sixty-eight flights).

In the early twentieth century, this hillside area had three times as many people living within its boundaries. Many of these residents were skilled and unskilled laborers attracted by employment opportunities in the glass and steel industries. Walking the South Side Slopes neighborhood offers a glimpse into the lives of those who lived along the hillsides perched above the mills. The houses, typically one room wide and several stories high, are packed together on narrow, winding streets connected by city steps and walkways.

The South Side Slopes Neighborhood Association has long advocated for neighborhood unity and safety while championing the improvement of public greenspaces and housing. Its annual October event, Step Trek, grew out of its organizational mission and offers self-guided tours of the steps and neighborhood. Former Pittsburgh mayor Tom Murphy (1994–2006) led five hundred people on the first Step Trek tour in 2000, and the event has continued to grow and attract more and more attendees every year.

BOUNDARIES: Josephine Street to the north, Sumner Street to the east, Arlington Avenue to the south and Fernleaf Street to the west.

DISTANCE: Approximately 3.0 miles.

DIFFICULTY: Moderate to challenging. Significant climb in elevation and many flights of stairs. Sterling Street has 301 steps.

PARKING: Street and metered parking is available throughout the South Side Flats. Consider parking along Josephine Street near the intersection with Greeley Street.

PUBLIC TRANSIT: The 48 bus connects to Josephine Street and South 26th Street. The 51 and 54 connect to Josephine Street and South 18th Street.

This walk travels along streets and stairs appearing in the South Side Slopes Neighborhood Association's Step Trek. For more information, see the "Resources" at the end of this book.

Start at the intersection of Josephine and Greeley Streets. This spot offers parking and nearby public transit options. Turn right on Greeley and take an immediate left onto Eola Way. Climb the Eola Way stairs (93 steps) (1) and notice several "orphan houses" on the left. Constructed during a time when few working-class families had cars, these homes are only accessible by the city steps and do not have immediate street access. While many such houses in other parts of the city have been abandoned and demolished over the decades, several in the South Side Slopes are well preserved.

After reaching the top, continue straight on Sterling Street. The first intersection will be Mission Street. To the left is the former St. Josaphat's Church and School. Built between 1909 and 1916, the church was named for Josaphat Kuntsevych, a bishop and martyr born in Poland in 1580. The building's design is credited to John Theodore Comes, a Pittsburgh-based architect. Comes was responsible for designing several Roman Catholic buildings throughout the country, including two in Pittsburgh's Lower Lawrenceville neighborhood: St. Augustine Church on 37th Street and St. John the Baptist Church on Liberty Avenue. A historic landmark, St.

City Steps Supporters and Enthusiasts

Over the last twenty years, the city steps have gradually gained recognition and support from neighborhood associations and enthusiastic visitors from around the country. Civic events featuring the steps have done much to call attention to their presence and aid in preservation efforts. Indeed, that was one of the main goals of the Fineview Citizens Council's Step Challenge race that began in 1995. With five-mile and two-mile race options, twelve of the neighborhood's public staircases are included for a total of more than 1,600 steps and breathtaking views. Because of its many years of support for the city steps, Fineview has received grants to create the Fineview Fitness Trail throughout the neighborhood, which features detailed mapping, signage and trail markers. To learn more about components of the Step Challenge and Fitness Trail, see the Fineview walking tour.

Since 2000, the South Side Slopes Neighborhood Association has conducted a similar event, the South Side Slopes Neighborhood Association Step Trek. Step Trek is not a race but a casual, day-long exploration of the many hillside steps, streets, parks and views the neighborhood offers. This fundraising event attracts hundreds of residents and visitors each year and offers participants two routes, aptly named "Black" and "Gold," after the official colors of the city and its professional sports teams. The organizers vary the Step Trek routes annually, but the steps count ranges between 1,200 and 1,800. A highlight of the walk is the "Big Bell Climb," which features a small tower bell at the summit of the longest flight that trekkers can ring to celebrate their arrival after they've caught their breath.

In addition to these long-standing annual events, you can find formal tours and informal explorations of city steps in various neighborhoods through social media groups and on activity websites such as MeetUp, as well as among local urban hiking and running groups and private businesses and tour operators in the city.

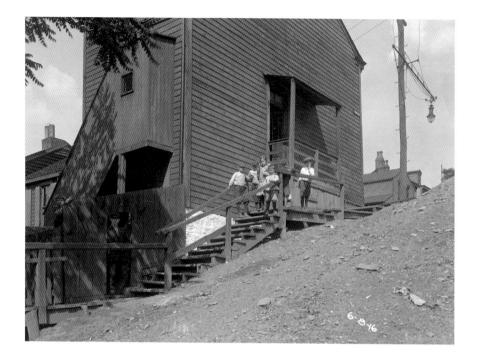

Josaphat's Church closed in the early 1990s. The building was sold and is now under private ownership.

Continue up Sterling Street using the sidewalk steps on the right (2). Surprisingly, this very narrow street, on which people park their cars, allows two-way traffic. As you walk up the 301 steps, you may witness vehicles reversing directions or trying to squeeze by each other. As you travel, take note of the three sets of wooden steps located on the left side. All provide access to Eleanor Street, which runs parallel to Sterling. The first is Caesar Way (52 steps) (3), the second is Holt Street (47 steps) (4) and the third is Sierra Street (52 steps) (5). Sierra features quite a few homes of newer construction, and almost all have back porches looking out over the expansive view.

Continue up Sterling until the "No Outlet" sign appears, which won't stop you on foot. At this point, the street ends, but the city steps continue. Don't forget to stop on occasion and turn to look back down Sterling to see how far and high you've climbed. If you're feeling a bit winded, it's to be expected. You'll have a view of the Birmingham Bridge (built in 1977 to replace the 1896 Brady Street bridge) stretched out below, which connects West Oakland to the South Side Flats.

To the right of the stairs is an entrance to the Sterling Connector Trail that links the city steps to South Side Park, a beautiful sixty-five-acre greenspace

Opposite: Today, Eola Way's city steps are made of concrete, but in 1916, the flight was made of wood and was called Lebanon Street.

Right: The 2023 view of the South Side Flats and Birmingham Bridge from the top of Eola Way.

in the heart of the Slopes neighborhood. Several trails within the park connect hikers to the community garden near Mission Street, athletic fields and a small playground near Saber Way/South 18th Street and the Arlington Baseball Field near Julia Street/Arlington Avenue. Signage and maps are located at each trail entrance and parking area.

When you reach the intersection for Salisbury Street, take a right. A sign for the St. Clair Incline is within half a block on the right. Operated by the St. Clair Incline Plane Company, this freight and passenger incline connected Josephine Street near South 22nd Street to Salisbury Street from 1886 through 1932. On April 6, 1909, at 3:30 a.m., St. Clair's incline engineer fainted and released the car's brake, causing it to roll back down the hill. Workers coming off the night shift from Cunningham Glass on Jane Street were trapped inside. One jumped from the car and survived; the others died from their injuries—a chilling reminder of the tragedy that once befell this epic Pittsburgh landmark. A South Side Park Trailhead connects the marker of the incline's location to the Sterling Trail below.

Turn left on Fernleaf Street and continue to the Arlington Spray Park and Playground. Take note of the 1894 Old Engine Firehouse No. 22 at the corner of Eccles and Fernleaf Streets. This corner held the firehouse, a school building on the opposite corner and the public park across the street. Of the three, the park, constructed by the city in 1921, has sustained continuous operation. The grounds are entered by stairs on Fernleaf or at the corner of Fort Hill Street (6).

Fort Laughlin, also known as Fort McKinley and Fort Ormsby, once occupied the playground's land. The circular, earthen structure, one of a network of thirty-seven similar structures, was built in 1863 by employees of Jones and Laughlin Iron Works (the precursor to J&L Steel) in response to a threatened invasion of Pennsylvania by the Confederate army during the Civil War. It was razed in 1924 to make room for the Arlington Playground and baseball field, and no evidence of the structure remains.

Exit the park by taking a left on Fort Hill Street and a right onto Sterling. From Sterling, turn left onto Arlington Avenue. After three blocks, take a left on Clover Street and descend the city steps (48 steps) (7). The first flight leads to Eccles Street, and the second flight (48 steps) connects to Baltic Way. Continue along Clover Street, take a right on Patterson Street and then left to climb up the Cologne Street sidewalk stairs (80 steps) (8). At the intersection of Berg Street, sidewalk steps are on both sides of the street (98 steps).

At Cobden Street, take a right and then a left on Sumner Street. Keep your eyes ahead for the spectacular view of the downtown Pittsburgh skyline. To the right of the street, the sidewalk dips below street level, and short wooden flights of stairs connect the street to each house. Continue to the end of Sumner and descend the city steps to Holt Street (110 steps) (9). This flight has an extension that once connected to Stromberg Street. Cross the street to access the sidewalk steps (58 steps) (10) and travel downhill to the intersection with Mission Street. Turn right and climb the 86 steps (11). Turn left on Oakley Street and descend the stairs to Stella Street and Shelly Street (98 steps) (12).

Descending the Sumner Street steps in the South Side Slopes.

Oakley Street is one of Pittsburgh's longest flights of freestanding stairs, with 285 steps (see the "Six Longest Flights" list). With endpoints at Josephine Street and Sumner Street, it has six cross-street intersections. This walk covers the lower half of this long and meandering flight. The upper half of Oakley Street features several flights of sidewalk stairs alongside streets allowing vehicle access.

At Shelly Street, look for the retaining wall. Near it, another flight descends to McCord Street (46 steps). The final leg of the Oakley Street stairs (74 steps) connects McCord to Josephine Street and features a mosaic mural built by Pittsburgh artist Laura Jean McLaughlin and one hundred community volunteers. Completed in 2016, this may be one of Pittsburgh's most photographed flights of city steps.

Take a left on Josephine and enjoy a well-deserved walk along level ground for 0.25 miles back to Greeley Street. The latter half features the freight rail line, constructed in 1873 for the Pittsburgh, Virginia & Charleston Railroad. Feel a sense of accomplishment in following the footsteps of the many generations of Pittsburghers who traveled the Slopes every day between work and home.

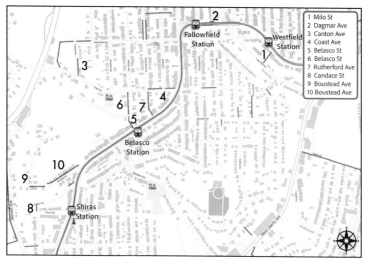

1	Milo St
2	Dagmar Ave
3	Canton Ave
4	Coast Ave
5	Belasco St
6	Belasco St
7	Rutherford Ave
8	Candace St
9	Boustead Ave
10	Boustead Ave

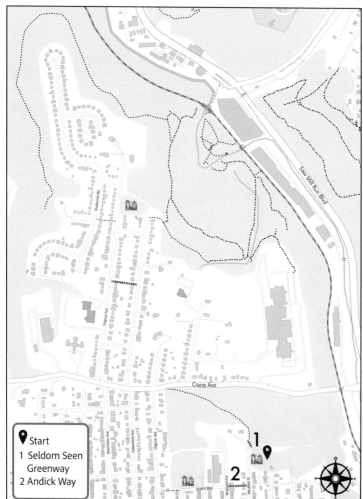

Start
1 Seldom Seen Greenway
2 Andick Way

BEECHVIEW AND BEECHWOOD

TRAVELING THROUGH TIME

Beechview's proximity to downtown Pittsburgh, convenient access to light rail transit, sweeping vistas and bustling main street make it an attractive place to visit. Located in the South Hills section of Pittsburgh, Scotch-Irish and English immigrants came to the area in the late 1700s and established the village of Orvilla. In the early years of the twentieth century, before the City of Pittsburgh annexed the area, this region's townships and boroughs were in a state of flux. The western section, known as Beechview, became its own borough after leaving Union Township. The eastern section, known as Beechwood, was part of West Liberty Borough. In February 1907, West Liberty Borough voted almost unanimously to join the city of Pittsburgh, and on January 4, 1909, the Borough of Beechview did the same. However, Beechview was selected as the official neighborhood name, with Beechwood relegated to history.

Today, Beechview's boundaries are Banksville Road, Saw Mill Run Boulevard, West Liberty Avenue and Wenzell Avenue. Beechview's main street, Broadway Avenue, laid with trolley tracks, brought commercial and residential development to the area. Those tracks remain, and today it is one of the few remaining places in Pittsburgh where vehicle traffic shares a portion of the road with light rail transit. The two tours here will enable you to explore today's Beechview and yesterday's Beechwood. One can be accessed by car or bus, while the other uses the Pittsburgh Regional Transit Light Rail System's Red Line.

TODAY'S BEECHVIEW CITY STEPS TOUR RIDING PITTSBURGH REGIONAL TRANSIT (PRT)

Beechview is served by the Dawn, Westfield, Fallowfield, Hampshire, Belasco and Shiras Pittsburgh Light Rail Red Line stations, part of the Port Authority of Allegheny County's light rail network. The Beechview stations serve a densely populated residential area through which bus service is limited because of hilly terrain. The Red Line, which travels along Broadway Avenue, is part of the line that runs between South Hills Village and downtown Pittsburgh. The stops listed here are within walking distance of sets of city stairways and steep residential streets. For light rail maps, timetables and fares, visit www.rideprt.org.

WESTFIELD STATION (0.1 MILES): This station is on Suburban Avenue on a raised platform directly across from the Milo Street city steps (96 steps) (1). This flight leads down to Cape May Avenue and West Liberty Avenue, which is less than a mile from the Liberty Tunnel (also known as the Liberty Tubes). It's difficult to imagine when traveling from downtown Pittsburgh to the South Hills didn't involve the Liberty Bridge and Tunnel. While trolleys connected Brookline and Beechwood as convenient public transportation, residents in this area needed to undertake the time-consuming task of navigating around or over Mount Washington or use its inclines to reach the South Side and all points east and northeast. Opening in 1924, the Liberty Tunnels are the longest automobile tunnels in Pittsburgh. The Liberty Bridge was then completed and linked to the tunnels on March 27, 1928.

FALLOWFIELD STATION (0.5 MILES): For those wanting to travel on foot or by bike to the Beechwood walking tour, this station is where you disembark. Climb down the city steps to Dagmar Avenue (76 steps) (2) and take a left. After crossing Sebring Avenue, Andick Way is on the right. You can complete the tour loop using that as your start and end point. The round-trip walk from this point is approximately 0.5 miles.

In 1912, Fallowfield Street was home to Engine Company No. 60. Because of the street's steep grade, the community challenged the city to provide the neighborhood with a better firefighting solution, as horse-drawn water tanks were ineffective. This public outcry led to Pittsburgh's first internal combustion motorized fire truck in Beechview. Today, Pittsburgh's Fire Station 28 is located a few blocks off Fallowfield at the corner of Sebring and Beechview Avenues.

HAMPSHIRE STATION (0.6 MILES): Overall, the streets in Beechview follow a grid pattern, but the neighborhood's hilly terrain results in some extremely

The Milo Street city steps, with their tangle of vines and fallen leaves, are viewed from Suburban Avenue.

steep roads. This includes Canton Avenue, the steepest street in the United States, with a 37.45 percent grade. The Canton Avenue sidewalk steps (54 steps) (3) are included in Pittsburgh's "Steepest Streets" list. You'll also find your way there by disembarking at Hampshire Station and taking a left (heading west) on Hampshire Avenue. After traveling six blocks down this hilly residential roadway featuring a mix of red brick and asphalt-paved

Above: The T stops near the top of the Belasco Avenue city steps.

Opposite: A deer crosses Boustead Street between the sets of sidewalk steps that line both sides.

streets, a dead end appears. The Canton Avenue sidewalk steps are on the left, and the Coast Avenue city steps (35 steps) (4) are directly ahead. A walker can loop around both, descending one while climbing the other. For those choosing to return to Broadway Avenue by Coast Avenue, it's important to note that this street does not have sidewalks.

BELASCO STATION (0.2 MILES): There are two flights of city steps very close to this station. To make a quick loop, travel down the Belasco Street stairs (93 steps) (5) and then continue down a long flight of sidewalk stairs (80 steps) (6)

to Coast Avenue. Take a right and, exercising caution, as the street is without sidewalks, travel for one block, turning right on Rutherford Avenue. Climb the long set of sidewalk steps and the standalone flight (123 steps) (7) back to Broadway. Looking to the right, you'll see the station stop. For those wanting to explore the area a bit more, the St. Catherine of Siena Roman Catholic Church is located at 1810 Belasco Avenue. The church offers services in English, Spanish and Portuguese as part of the Saint Teresa of Kolkata Parish. The Beechview branch of the Carnegie Library of Pittsburgh is across the street at 1910 Broadway Avenue. This location was originally home to the Beth El Synagogue, which opened in 1927. The synagogue moved, and the property was sold to the City of Pittsburgh in 1965. The library opened in 1967 as the seventeenth branch in the Carnegie Library of Pittsburgh system.

SHIRAS STATION (0.6 MILES): A few steps from the station is the Beechview Monument Parklet at Broadway and Shiras Avenues. Memorials honor soldiers from World Wars I and II, the Korean and Vietnam Wars and veterans who have served our country worldwide. The parklet offers benches and paved walkways between the memorials. In honor of Beechview's centennial in 2005, a time capsule was created and placed in the park. The capsule will be opened in July 2055. To visit nearby city steps, cross Broadway Avenue heading away from the parklet and walk down Shiras Avenue. The flight (56 steps) (8) is on your right. Descend to Crosby Avenue. Turn left on Crosby and right onto Wenzell Avenue, crossing the street to access sidewalks. In one block, turn right on Boustead Avenue, which has a 29 percent grade. Flights of old (59 steps) (9) and new (76 steps) sidewalk steps (10) line the street but are not continuous, meaning it is necessary to cross the street at different points. Once you reach the top, turn right on Broadway, and you'll soon be back at the station.

YESTERDAY'S BEECHWOOD
AND THE SELDOM SEEN GREENWAY

BOUNDARIES: Orangewood Avenue to the east, Crane Avenue to the north, Dagmar Avenue to the west and Andick Way to the south.

DISTANCE: Approximately 1.0 mile.

DIFFICULTY: Moderate. The walk fluctuates between uphill and downhill and involves a dirt trail through Seldom Seen Greenway. Dagmar Avenue is very steep but offers sidewalks.

PARKING: A parking lot is available at Vannucci Field and Beechview spray park on Orangewood Avenue.

PUBLIC TRANSIT: Due to the steepness of Crane Avenue, bus transportation to this area is limited. The 36 bus stops at Banksville Road and Crane. From there, it is a half-mile walk uphill with limited sidewalk access to Dagmar Avenue.

Since it was formally established in 1985, Beechwood's Seldom Seen Greenway has lived up to its name as a quiet and infrequently visited hillside area. The valley once held an isolated village tucked between the railroad tracks and the Saw Mill Run, which was accessible only through a tunnel beneath the railroad. After it was annexed to the city of Pittsburgh in 1924, residents dwindled, with the last leaving in the 1960s.

While a large portion of this greenspace is sandwiched between Banksville Road and Saw Mill Run Boulevard, a smaller portion is south of Crane Avenue and adjacent to the Beechview spray park and Vannucci Field. With ample parking, the park is an ideal place to start this walk. While the Seldom Seen trail in this section is only a quarter of a mile, it has a fifty-foot elevation change, and a new flight of wooden city steps (17 steps, no

Top: A new set of steps leads into the Seldom Seen Greenway. *Bottom*: Restoration of native plants helps to preserve the hillsides.

PITTSBURGH'S NEIGHBORHOOD PLAYGROUNDS

The first public playground in the city of Pittsburgh was built in 1896 and located in what is now downtown, at Second Avenue and Grant Street. Soon, neighborhoods around the city began requesting playgrounds. A corner lot with a swing set, slide, sandbox, small shelter and some open space was common. Some of the nicer parks included a ballfield or a basketball hoop. Throughout the 1920s, the city initiated a program to upgrade existing playgrounds and construct public parks with recreation centers, pools, ballfields and other facilities. The financial strain of the Great Depression slowed this activity, but the Civilian Conservation Corps and the Works Progress Administration put the Pittsburgh parks program back on track in 1935. Today, the city has 165 parks scattered throughout all ninety neighborhoods, offering amenities such as playgrounds, spray parks, public swimming pools, picnic areas, walking trails, river access and athletic courts and fields.

railing) (1) makes a considerable difference in accessibility. Signs marking the entrance to the Seldom Seen Greenway trail are located at the north end across from the spray park.

The Seldom Seen trail meanders through the woods to Crane Avenue. Historical maps of the area indicate that the Theodore Lau (Low) family owned this land dating back to 1890, before it was a part of Pittsburgh. Several structures existed on their expansive property, and today's nature trail appears to follow the original access road.

To explore the rest of the Greenway, turn left on Crane and then take your second right onto Gladys Avenue. This will lead to Tropical Park and additional trails. To continue with the Beechwood walk, turn left on Crane and then another left onto Dagmar Avenue. With a 20 percent grade, Dagmar is one of Pittsburgh's steeper streets, and while walking up is a workout, it's considerably less nerve-wracking than driving in a vehicle. Once you reach the summit, turn left on Andick Way and follow this city-owned paved walkway past the Alton Playground and downhill to Rockland Avenue. The Beechwood Elementary School will be to the left. Before the annexation to Pittsburgh, Beechview and Beechwood were adjacent neighborhoods. While the area around Crane Avenue and the Seldom Seen Greenway is now

The Andick Way city steps as viewed from Westfield Street.

called Beechview, the old Beechwood name lives on in the neighborhood's public school title. Architect Press C. Dowler designed this school, which opened in 1924. It was added to the National Register of Historic Places in 2002. Theodore Lau, son of the original landowner Theodore Lau, likely walked Andick Way to attend the Beechwood Public School, which featured a public playground next door.

Continue down the Andick Way city steps (38 steps) (2). This leads to Orangewood Avenue, Vannucci Field's entrance and the Beechview spray park. While the playgrounds of the early twentieth century were quite a bit different from today's, the sounds of children playing have largely stayed the same.

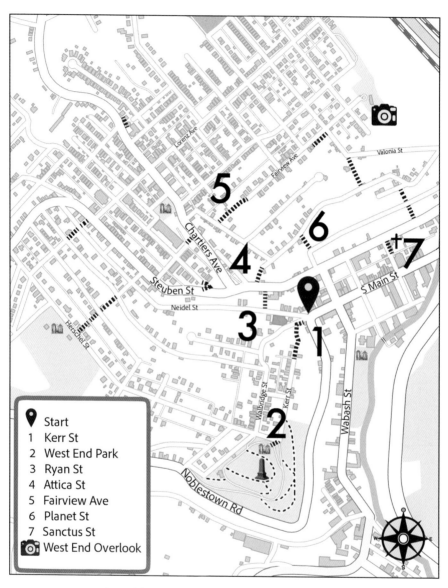

Start
1 Kerr St
2 West End Park
3 Ryan St
4 Attica St
5 Fairview Ave
6 Planet St
7 Sanctus St
West End Overlook

openstreetmap.org/copyright

WEST END VILLAGE AND ELLIOTT

HISTORY AND SCENIC VIEWS

Just downstream of the confluence of Pittsburgh's rivers is an area of steep, tree-covered hills with some of the best views the city has to offer. West End Village and Elliott neighborhoods have a quiet feel with several parks, greenspaces, city steps and a healthy dose of intriguing historical tidbits—including how the neighborhoods came to be. Much of the area was owned by West Elliott, who built a home along Saw Mill Run in 1815. Unfortunately, by 1835, Elliott needed to sell most of his land to pay off gambling debts. The sale of that land resulted in Temperanceville (today's West End), founded as a dry town. All lots sold included a clause stipulating that no liquor could be sold on the parcel. Nevertheless, by the mid-1800s, several illegal saloons and other business enterprises associated with the coal mining industry flourished.

Temperanceville was annexed to the city of Pittsburgh in April 1872 and renamed West End Village. The neighborhood continued to grow, reaching nearly four thousand people by 1900. Slavs, Hungarians and Poles joined the English, Irish and Germans in the neighborhood, and soon it became a shopping, transportation and entertainment hub for western Pittsburgh. But frequent flooding in this low-lying area often turned the streets into rivers of mud after rain or snow.

The West End Bridge greatly enhanced access to West End Village and the many neighborhoods within the West End. Opening in 1932, this steel tied-arch bridge over the Ohio River connects the West End to the Chateau neighborhood on the North Side. It was the longest tied-arch bridge in the world when completed. It joined the National Register of Historic Places in 1979 and became a historic landmark in 2001.

BOUNDARIES: West End Park to the south, West End Overlook to the north, Sanctus Street to the east and Walbridge Street to the west.

DISTANCE: Approximately 3.5 miles.

DIFFICULTY: Light to moderate. Flights of stairs with one hundred steps or more and some narrow streets without sidewalks.

PARKING: Street parking is easily found on South Main Street and is available at both West End Park and the West End Overlook.

PUBLIC TRANSIT: The 31 and 27 bus will take you to Steuben Street.

West End Village is in a valley separated from neighboring Elliott and Duquesne Heights by steep hillsides. The city steps in this neighborhood are pretty impressive (including the ones closed for safety reasons), and walking them offers dramatic views of the area. This walk begins on South Main Street at the intersection of Walbridge Street (1). The Kerr Street city steps are to the left. As you walk up these 199 steps, take a moment to pause and enjoy the hillside and skyline views. Next, continue to walk along the red brick–paved Kerr Street to the intersection of Herschel Street. A flight of stone stairs leads into West End Park (2). To access the park's formal entrance, pavilion and playground, continue on Herschel Street and turn left on Walbridge Street, lined with Belgian block. The park includes a World War I monument honoring the members of the 16th Military Zone by Frank Vittor, the prolific sculptor responsible for many other statues and memorials throughout the city. After exploring the park, retrace your journey back down Kerr Street and the stairs.

Turn left onto South Main Street and look for the flight of city steps (48 steps) across the street from the massive stone church (3). Designed in 1887 by the architectural firm Longfellow, Alden & Harlow as the West End United Methodist Church, the building has a slate gable roof with arched leaded glass lunettes and ribbon windows. The church's other features include a corner tower with a pyramid slate roof and an arcaded open belfry. The entranceway is a large semicircular arch with a date stone above in a corner. The building is now home to the House of Transformation

RUNNING CITY STEPS WITH NEBBY N'AT

The word *nebby* is decidedly Pittsburgh. While its general meaning is "nosy," it can be used to describe behavior ranging from inquisitive to annoying and is the perfect word to describe anyone seriously committed to exploring the city steps. Tyler Abernathy and Chris Richards—along with a few dozen other men of every age, background and physical ability level—meet up most mornings at 5:30 a.m. to run stairs and hillsides throughout all areas of the city with the Nebby N'at running group. Abernathy and Richards are transplants from South Carolina who met through the free men's exercise group F3, which promotes fitness, fellowship and faith.

"There's no better way to see the city than on foot," Abernathy says. "The first time I ran in the South Side Slopes, we were on the Sterling Street city steps, and every time I thought I was done, I realized, 'Oh wait, there's still more!'" Richards, who launched the group on January 1, 2022, echoes that sentiment. "Exploring the city this way is so different from driving," he explains. "It's always a new experience, and when we hit the stairs, all that mumble chatter that often goes through your mind when you're exercising takes on a whole new dimension. When new members see the stairs for the first time, they either love it or we never see them again!"

Nebby N'at regular Ryan Hoover agrees. He ventured out with the group for the first time on a Fineview run that involved running up and down Rising Main, one of the longest flights in the city—not once, but twice. "It was pretty crazy," Hoover admits, looking back on the challenge, "and I wondered for a moment what I had gotten myself into, but I made it through." One of the group's favorite locations is running through the hills in Elliott to the West End Overlook. "There's such a great feeling of looking out from that beautiful spot and seeing the sunrise," Hoover says. "It's an amazing way to experience Pittsburgh."

Global Ministries, which offers religious and community-building services to African Christians.

At the top of the city steps is Steuben Street. Cross the street to climb the Attica Street stairs (90 steps) (4). Take a left on Furley Street and a right onto the Fairview Avenue stairs (101 steps) (5). At the intersection of Marena Street, the sidewalk dips below street level, and short flights of stairs connect the houses below to the street above. At Valonia Street, the sidewalk

Opposite: The Planet Street city steps climb a steep hillside in the West End.

Above: Today, the West End's Balfour Street city steps travel through an urban landscape surprisingly rural in appearance.

reemerges to street level. Continue on Fairview to the entrance to the West End Overlook Park.

While many consider Grandview Avenue in Mount Washington to have the best skyline viewing area in the city, West End Overlook Park is a very close second. Perched seven hundred feet above the Ohio River, the park offers a shelter house, parking lots, ADA access and paved walking paths.

After marveling at the view, walk back down Fairview Avenue and take the third left onto Valonia Street. At the end of the street is a flight of city steps leading to orphan houses halfway down the hillside. This flight once continued to Balfour Street, but the bottom portion is currently closed. Follow the detour down Advent Street, take a right on Angle Street and then another right on Balfour Street. A glance down Lander Street reveals a closed sign on another segment of the Balfour Street stairs leading downhill. Instead, walk uphill along Balfour, and on the right, the bottom portion of the flight leading to the orphan house appears. Continue along Balfour until it turns into Attica Street. The Planet Street stairs (133 steps) (6) are on the left. When reaching the bottom, take a left on Elliott Street and continue

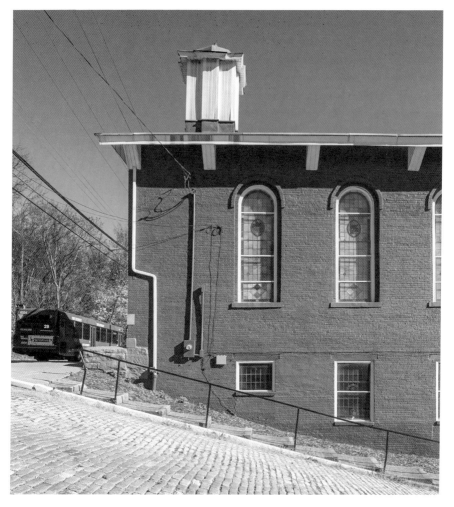

Sanctus Street is short but steep, and the sidewalk steps make walking a bit easier.

walking until you reach the Balfour Street stairs again. After pinballing back and forth on these hillside switchback roads, it's time to head down.

Take a right on Steuben and a left on Sanctus Street. The former German Evangelical Church (now Jerusalem Baptist Church) is at the corner. Built in 1864, the church is a historic landmark. Sanctus Street features a flight of sidewalk stairs on the left (31 steps) (7) and a Belgian block street. Continue down to South Main Street. At the end of the block is St. James Church, a Roman Catholic Gothic-style church built in 1884. The church served the community for 120 years until its closure in 2004. Today, the Society of

Saint Pius X oversees the church and offers the traditional Latin Mass to the Pittsburgh community.

Take a right on South Main Street and walk through the central business district, which features coffee shops, art galleries and restaurants. For a quick detour, take a left on Wabash Street and travel one block to the West End branch of the Carnegie Public Library, a historic landmark. This branch opened in 1899 and is one of the original Carnegie Library buildings. This location is also home to the very first children's story time in Pittsburgh libraries. Started in 1900 by Francis Jenkins Olcott, story time involved reading nursery tales aloud to young children to help prepare them for learning how to read. Inside is a small local history collection, including books, photos and other materials directly related to the West End neighborhood and its history.

Along with its historical collection, the library houses dozens of handmade models of buildings in the neighborhood. Replicas include homes, businesses, churches and, of course, the library—all of which are on permanent display inside. Once you've finished touring the library, retrace your travels along Wabash Street and take a left on South Main Street, and the Kerr Street stairs will come into view.

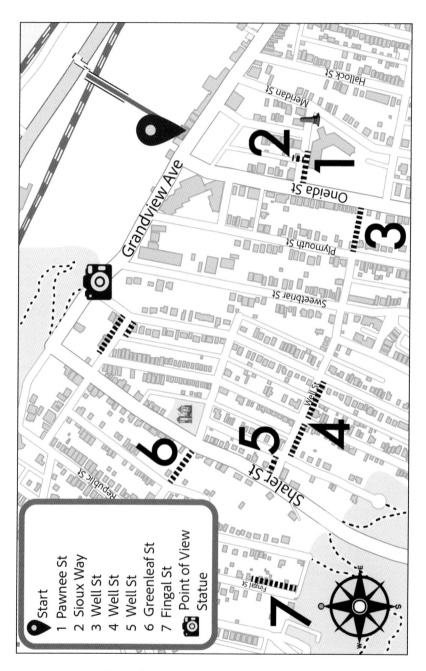

Start
1 Pawnee St
2 Sioux Way
3 Well St
4 Well St
5 Well St
6 Greenleaf St
7 Fingal St
Point of View
Statue

DUQUESNE HEIGHTS

CAPTIVATING POINTS OF VIEW

In 1770, George Washington and Guyasuta, the Native American leader of the Seneca people, traveled the six-hundred-foot hillside that slopes below Grandview Avenue. Little did they know that within one hundred years, the area would become a key site in Pittsburgh's industrial age, with workers mining some 13 million tons of bituminous coal every year. While the entire hilltop area, which includes the adjacent neighborhood of Mount Washington, would come to be known as "Coal Hill," the namesake of Duquesne Heights is Fort Duquesne, the French outpost constructed in the 1750s. Duquesne Heights, originally known as "Dutch Hill," was annexed into the city of Pittsburgh in 1872, after which many city steps were constructed along the hillsides, aiding residents and workers traveling to and from industrial areas along the Ohio and Monongahela Rivers.

In 2006, a community effort brought together three separate greenspaces—Grandview Avenue, Grandview Park and Olympia Park—into one unified park named Emerald View. One of many regional parks in Pittsburgh, Emerald View boasts winding trails, wooded hillsides, scenic overlooks and sweeping skylines. It also offers ample opportunities to observe wild birds along the open river flyways and hillside woodlands. For this picturesque walk, consider packing binoculars.

BOUNDARIES: Grandview Avenue to the east, Greenleaf Trail to the north, Well Street to the south and Fingal Street to the west.

DISTANCE: Approximately 4.5 miles, including the Greenleaf Trail and Emerald View Trail. The walk has a street side–only route that is 2.5 miles.

DIFFICULTY: Moderate to advanced. The street side walk has uphill climbs. The Greenleaf Trail and Emerald View Trails are not paved and may be muddy or rough in spots.

PARKING: Street parking is easily found near Grandview Avenue. A parking lot is available on West Carson Street for those wanting to ride the Duquesne Incline.

PUBLIC TRANSIT: The 40 bus services Grandview Avenue and Shaler Street in Duquesne Heights.

This walk begins at the Duquesne Incline on Grandview Avenue. Two of the four inclines that served Mount Washington still exist: the Monongahela Incline, which opened in 1870, and the Duquesne Incline, which opened in 1877. The route of the Duquesne Incline follows the tracks of an early coal hoist from 1854.

The Duquesne Inclined Plane Company operated its incline until 1962. The Society for the Preservation of the Duquesne Heights Incline took over in 1964. Travelers marvel at the incline's original red wooden cable cars, and the upper station on Grandview Avenue houses a museum of Pittsburgh history, including photos and information about inclines from around the world. In the gift shop, you'll find Pittsburgh souvenirs, maps and photos.

After exiting the incline, turn left on Grandview Avenue and right on Meridan Street. At the intersection of Pawnee Street, turn right. You'll see a World War I memorial dedicated to neighborhood residents who served at the intersection. While Pawnee Street ends, its 52 city steps follow alongside Whittier Elementary School (1). The school's name honors Isaac Whittier, an early director of Mount Washington's schools before its annexation to the city of Pittsburgh in 1872. Designed by M.M. Steen, the building opened in 1939 and still serves Mount Washington and Duquesne Heights.

As you walk down the stairs, a short flight (8 steps) intersects at Sioux Way (2). Travel down to admire the original red brick street. You may notice how the part of Sioux Way connecting to Oneida Street is paved with asphalt. Based on the 1923 street map of the neighborhood, Sioux Way ran parallel to Oneida and Cohasset Streets with endpoints at Ponka Way and the Pawnee Street stairs. The asphalt connection followed the removal of homes once lining the street. The same maps also show that Holliday Park once occupied the land where the school now stands, and the park's stone slab steps accessible from Oneida Street remain in excellent condition.

A 1935 view of the Indian Trail steps as they climbed the dangerous hillside to Mount Washington. *Pittsburgh City Photographer Collection, University of Pittsburgh.*

Take a left on Oneida and cross the street to descend the Well Street city steps (81 steps) on the right (3). Continue along Well Street for ten blocks. Old railings on the sides of the street make the remains of old sidewalk stairs more noticeable. Today, nature is reclaiming these sections of sidewalks. However, Well Street is narrow, so you may prefer the sidewalk. After crossing Edith Street, the stairs are fully functional (48 steps) (4) and easy to travel. It's worth crossing to the other side of the street to get a view of the old red brick water runoff channel (see the "Rebuilding City Steps" chapter). At the intersection of Clarence Street, cross to the other side and take the last segment of city steps to Shaler Street (26 steps) (5). Turn right on Shaler and continue to Greenleaf Street. To the right is the Eileen McCoy playground, and to the left is a flight of wooden city steps (74 steps) (6). Climb the stairs and continue along Greenleaf Street, which features old sidewalk steps on both sides and another red brick water runoff channel.

Take a left onto Fingal Street and look for the sidewalk stairs below street level (26 steps) (7) on the left. Fingal loops around to join Republic Street, and an entrance to the Emerald View Park's Duquesne Heights Greenway is on the right. Follow the directions below to hike the Greenleaf Trail

Opposite: The Well Street city steps viewed from Oneida Street. Tucked between houses, this flight can be easy to miss.

Right: The Fingal Street sidewalk steps feature a mix of construction materials including concrete, brick and wood.

and the Emerald View Trail. To bypass the trails and stay on residential streets, continue walking on Republic Street for three blocks until it ends at Grandview Avenue. Enjoy the "grand views" and continue until you reach the *Point of View* sculpture and monument. Within two blocks, the Duquesne Incline comes into view.

GREENLEAF TRAIL AND EMERALD VIEW TRAIL

Enter the park through the fenced trail entrance. This section winds you through lovely woodlands and over natural streams for about 0.5 mile, where you'll see a trailhead for one of Greenleaf's parking areas. Next, head down a wooded switchback filled with sweeping views of West End

This page: A modern view from the Emerald View Trail where the Indian Trail city steps once stood (1934). *Pittsburgh City Photographer Collection, University of Pittsburgh.*

Village. The trail crosses Greenleaf Street (use caution while crossing). Once across, head left up Horner Street to reenter Emerald View Park to walk the Emerald View Trail.

The Emerald View Trail traverses Mount Washington's hillside and provides magnificent views of downtown Pittsburgh, the West End, the North Shore and the Ohio, Allegheny and Monongahela Rivers. After about one mile, the trail emerges into an open restoration and viewing area. Take a rest on the sitting stones and enjoy the view. Then, proceed up the gravel path to Jim West's *Point of View* sculpture of George Washington and Seneca leader Guyasuta. Dedicated in October 2006, this oft-photographed piece of public art represents the famous meeting of the two leaders and is a focal point for the neighborhood and the region. The hillside below was the location of the legendary Indian Trail steps, which was more than one mile long with one thousand wooden steps. Keeping the rivers to the left, continue along Grandview Avenue for two blocks. The Duquesne Incline appears on the left.

EXPLORING THE SIX LONGEST FLIGHTS OF CITY STEPS

While long walks exploring neighborhoods and their city steps can be an enjoyable way to spend a few hours, at other times, a shorter but more intense experience is in order. Many flights in Pittsburgh are between one hundred and two hundred individual steps (even if they feel much longer mid-flight), but the longest flights greatly exceed that range. The six longest flights in the city offer opportunities to experience residential areas in the South Hills, the South Side Slopes, the North Side and the East End. Some of the flights listed here are included in longer neighborhood walks. In those instances, the walk is referenced below.

RAY AVENUE, BROOKLINE (378 STEPS, PARTIALLY CLOSED)

The Ray Avenue city steps in Brookline is the longest flight in the city and, for more than one hundred years, helped move people throughout this South Hills neighborhood bordered by West Liberty and Pioneer Avenues. With 378 individual steps, Ray Avenue first appears as a street on the 1910 city map and then as a flight of stairs in 1923. The concrete version we see today was built in 1954. Unfortunately, only the middle section between Plainview and Woodward Avenues is clear and open. The top and bottom segments have "Closed" signs because sections of concrete have fallen away, leaving gaps. However, visiting the middle section is worth it when exploring this longest flight.

JACOB STREET, OVERBROOK/BROOKLINE (364 STEPS)

The Jacob Street city steps, the second-longest flight in Pittsburgh, are split between the South Hills neighborhoods of Brookline and Overbrook. Unlike Ray Avenue, which holds the top spot, this flight is entirely traversable from start to finish in a unique, parklike setting. If driving to the area, park near the end of Sunbeam Way and follow a footpath to the first long flight of stairs. Once you reach the bottom, you'll find a concrete walkway, several packed dirt trails and the second segment of stairs that heads up to Brookline Boulevard. The trails lead to various places, including Brookline Memorial Park, the South Busway and the freight rail tracks (an ideal place for trainspotting, but exercise caution, as the area is unobstructed). It's a lovely spot and worthy of investigation any time of year.

57TH STREET, UPPER LAWRENCEVILLE/STANTON HEIGHTS (345 STEPS, PARTIALLY CLOSED)

The 57th Street stairs start at Christopher Street on the border of Upper Lawrenceville and Stanton Heights and are open through the lower half of its segment between Duncan and Camelia Streets. The street appears on maps dating back to 1890—with stairs indicated on maps in 1903, 1910 and 1923—and presumably was built to lead to the Mount Albion/Sunnyside Public School once located on McCandless Avenue. While walkers can no longer fully explore this flight, it's worth visiting to experience an example of orphan houses. In today's Pittsburgh, many homes without street access are now abandoned and often demolished. However, 57th Street still has a few. Perhaps in time, the growing popularity of this neighborhood will result in improved access and new life for old buildings.

RISING MAIN, FINEVIEW (331 STEPS)

The Rising Main city steps in the North Side neighborhood of Fineview is the fourth-longest flight in Pittsburgh with 331 steps. The flight is best accessed from Howard Street, a quiet road parallel to Highway 279. Walking along Howard from East North Avenue takes you past two other

This page: The Jacob Street city steps connect Brookline and Overbrook to the South Hills T line.

The 57th Street steps in Upper Lawrenceville once provided a connection to McCandless Avenue in Stanton Heights.

flights worth investigating: Habit Way (144 steps) and Carrie Street (106 steps). A public works building has occupied the base of Rising Main and Toboggan Way (78 steps) since the 1880s, with the stairs appearing on maps starting in 1902. The condition of the current flight is stable but has been significantly affected by leakage from underground water and sewer pipes, as well as hillside erosion exacerbated by the demolition of abandoned properties. Once you reach the top, you can continue to climb the wooden steps that line the paved portion of Rising Main Avenue (55 steps) and head toward the many city steps that line Lanark and Marsonia Streets. See the "Welcome to Fineview" walking tour for a longer trek.

Yard Way, South Side Slopes (317 steps)

The Southside Slopes neighborhood has the largest number of city steps, with sixty-eight individual flights carrying people up and down this densely populated hillside area. Located in the western section of the Slopes, Yard Way is the longest flight in the neighborhood, with 317 steps. This flight has endpoints at Pius Street (close to the Southside Flats) and Shamokin Street (near Winters Playground and Park) and comprises six segments that intersect with paved city streets. Each intersection features a street sign with a stairway graphic, indicating Yard Way as an official city street—albeit one that isn't accessible to vehicles. The South Side Slopes has many impressive flights, and any neighborhood exploration should include this one, as the views are stunning once you reach the top. See the "South Side Slopes West: Stairways to Heaven" walking tour to extend your trip.

Oakley Way, South Side Slopes (285 steps)

Located in the eastern section of the South Side Slopes, Oakley Way is the second-longest flight in the neighborhood with 285 steps. This flight has endpoints at South 27th Street (close to the Southside Flats) and Sumner Street. Access to Monongahela Park is available through the Oakley Way intersection with Stella Street. The flight combines free-standing flights of stairs and sidewalk steps alongside paved city streets. The lowest portion of the flight features a colorful mosaic mural designed by Pittsburgh artist

Laura Jean McLaughlin. Any neighborhood exploration should include this lower section of the flight, as it's one of Pittsburgh's most frequently photographed places. See the "South Side Slopes East: Sterling Efforts Yield Rewards" walking tour to challenge yourself.

PITTSBURGH'S STEEPEST STREETS

Pittsburgh contains some of the steepest streets in the United States. They're often a terrifying nightmare for drivers and a demanding physical challenge for cyclists, but for pedestrians traveling the city steps, they're a surprisingly surmountable challenge. The following list includes only those Pittsburgh streets that have a grade at or greater than 23 percent and a corresponding flight of public stairs.

These ten steepest streets offer opportunities to experience residential areas in the South Hills, the North Side and the East End. Some of the flights listed here are included in longer neighborhood walks throughout the book. In those instances, the walk is referenced below.

If driving or biking to these destinations, obey all traffic laws (some streets are one-way) and ensure that your brakes are in good working order before traversing downhill. Streets paved with Belgian blocks require extra precaution.

Canton Avenue
Grade: 37 percent
Steps: 54
Neighborhood: Beechview
Included in the "Beechview and Beechwood: Traveling through Time" walking tour.

Connecting Coast and Hampshire Avenues in the South Hills, Canton Avenue may be the steepest street in the United States, but it's only one block long. Canton is a one-way street paved in Belgian blocks; for motorists, all traffic travels uphill. When you reach the top, look to the left for the entrance of the Hampshire Avenue city steps (112 steps). This stairway connects to the Graymore Street stairs (35 steps) and creates a loop down to Coast Avenue. While safe to walk, the sidewalk portions of this flight are eroding because of the deteriorating hillside.

DORNBUSH STREET
Grade: 32 percent
Steps: 273
Neighborhood: East Hills

Dornbush is the steepest street in the East End and the second-steepest street in the city. All traffic flows downhill on this one-way paved roadway, which stretches for several blocks with cement stairs on the side for pedestrians. Some, but not all, sections of the stairs are clear and traversable.

BOUSTEAD STREET
Grade: 29 percent
Steps: 76
Neighborhood: Beechview
Included in the "Beechview and Beechwood: Traveling Through Time" walking tour.

Boustead Street is the second-steepest street in Beechview and the third steepest in the city. This flight is significantly longer than Canton Avenue and connects the residential area of Wenzell Avenue to the Broadway Avenue business corridor and the T (light rail). Vehicle traffic flows in both directions.

NANSEN STREET
Grade: 28 percent
Steps: 152
Neighborhood: Hazelwood
Included in the "Hazelwood: With Every Step, Another Story Unfolds" walking tour.

This steep East End street connects the hillside between Flowers and Hazelwood Avenues and features a mix of traditional sidewalk steps with catwalk-style platform stairs raised several feet above the roadway. In 2019, Nansen Street was closed to vehicles because of nearby landslides. The best approach for pedestrians is by way of Flowers Avenue.

Above: The sidewalk steps offer an easier way to travel up or down Canton Avenue, the steepest street in the continental United States.

Left: A view from the middle of Dornbush Street looking toward Bricelyn Street.

Above, left: This section of the Boustead Street sidewalk steps shows the street's steep grade.

Above, right: With its Belgian blocks, Cutler Street and its sidewalk steps offer a glimpse into neighborhood life from a different era.

Left: Rialto Street and its sidewalk steps from the Lowrie Street Bridge. The Allegheny River and 31st Street Bridge are in the distance.

EAST WOODFORD AVENUE
Grade: 27.6 percent
Steps: 61
Neighborhood: Carrick

The rolling hills of Carrick in Pittsburgh's South Hills are on full display when descending East Woodford Avenue. While this steep street begins at Brownsville Road, the city steps are at the intersection of Hazeldell Street and below street level. Arden Way, which runs parallel to East Woodford, provides access to the homes that line this section of the street.

CUTLER STREET
Grade: 26 percent
Steps: 106
Neighborhood: Perry South

Located in the North Side's Perry South neighborhood, Cutler Street connects Perrysville and Wilson Avenues. Cutler is only one block long, but traversing the Belgian blocks can be a challenge for motorists and cyclists. The sidewalk steps make for an easier walk and a pleasant way to admire this long-ago paving style, but sections may be affected by pervasive Japanese knotweed in the warmer months.

RIALTO STREET
Grade: 24 percent
Steps: 167
Neighborhood: Troy Hill
Included in the "Troy Hill: Where Everything Old Is New Again" walking tour.

For many Pittsburgh motorists, Rialto Street on the North Side's Troy Hill neighborhood is a nightmare. This narrow, two-way street can just barely accommodate vehicles passing each other and is frequently closed to traffic during inclement weather. Walking these stairs provides a front-row seat to the white knuckles and hot tempers that frequently flare. While the lower portion of Rialto connects to the Route 28/East Ohio Street bike/pedestrian trail, you can access the upper portion via Ley Street or the stairs from the Lowrie Street bridge.

NORTH WINEBIDDLE AVENUE
Grade: 23 percent
Steps: 154
Neighborhood: Garfield

North Winebiddle Avenue, which begins at the Penn Avenue business corridor in the East End neighborhood of Garfield, starts off relatively flat, but after a few blocks, the grade increases. Sidewalk city steps appear at Kincaid Street and continue two blocks to Rosetta Street, where the road technically ends. However, the "paper street" section of North Winebiddle continues to Hillcrest Street thanks to an additional flight of stairs. As the elevation increases, stunning views appear of downtown and Oakland, including, most notably, the University of Pittsburgh's forty-two-story Gothic Revival skyscraper known as the Cathedral of Learning.

HAMPSHIRE AVENUE
Grade: 23 percent
Steps: 147
Neighborhood: Beechview
Included in the "Beechview and Beechwood: Traveling through Time" walking tour.

Hampshire Avenue is the longest of the steep streets and has both hills and flat stretches during its twelve-block run. The western end of the street intersects with our steepest street, Canton Avenue. When you reach the end of Hampshire, look ahead for the entrance of the Hampshire Avenue city steps (112 steps). This stairway connects to the Graymore Street stairs (35 steps), which exit onto Coast Avenue.

NEWETT STREET
Grade: 23 percent
Steps: 24 on Newett, plus 196 on Copperfield Avenue
Neighborhood: Carrick

Newett Street, which intersects with Brownsville Road in the South Hills neighborhood of Carrick, may only have 24 sidewalk steps, but nearby Copperfield Avenue offers many more (196; see "Rebuilding City Steps"). Simply make a loop through this pleasant residential area, walking Newett, Mount Joseph Street, Copperfield and Brownsville for a challenging climb and rewarding descent.

OFF-ROAD ADVENTURES

Pittsburgh's steps were built when the city's population was nearing 700,000 residents. As the current population hovers around 300,000, it's no surprise that many greenways and wooded hillsides once contained businesses, homes, streets and alleys. As the population declined over several decades, Pittsburgh's natural world aggressively reasserted itself into spaces no longer inhabited or actively maintained. Although structures and streets disappeared into nature, the concrete and steel of city steps remain, themselves gradually falling into decay.

Unlike a guided walk along neighborhood streets, this section offers a collection of unique locations intended to stimulate curiosity and pique interest. Some locations are historically significant, while others are akin to "off-road" hiking and scrambling along hillsides. Each site is an opportunity to experience Pittsburgh in a unique way not found in many other urban centers. When visiting these locations, traveling alone is discouraged, as hillsides are steep and often without nearby residences. In addition, visiting these spots from late spring through early fall may disappoint, as plants' natural growth could obscure both the stairs and the ground. However, with a healthy dose of caution and preparation, the urban explorer can safely discover spots of natural beauty infused with Pittsburgh's unique brand of post-apocalyptic charm. As always, exercise extreme caution and care when traversing these routes.

ONE HUNDRED YEARS OF RUBBISH

Litter and trash along the stairs have always plagued the city steps, in some cases blocking pedestrian access and making for unsightly travel. Long before 311, Twitter, Reddit and social media, the "Mr. Fix-It" column in the *Pittsburgh Press* published residents' complaints and advocated for remediation at the appropriate city department. Familiar concerns throughout the 1930s included stairs and sidewalks needing repair, weed removal, illegal dumping on hillsides and trash and litter accumulating in empty lots.

In the one hundred years since those early complaints, the problem has escalated in neighborhoods experiencing significant depopulation. Empty hillsides and overgrown lots quickly attract tires, construction debris, unwanted furniture and bags filled with trash. Large items such as refrigerators, kitchen stoves, bathtubs, television sets and abandoned vehicles and small motorboats also appear regularly.

Since 2000, Allegheny CleanWays, a Pittsburgh-based nonprofit, has engaged and partnered with community groups to remove more than 5.4 million pounds of debris, including more than forty-five thousand tires, from vacant lots, wooded hillsides, alleyways, roadsides, streams and riverbanks. Each year, hundreds of volunteers and neighborhood groups work with them to remove trash from the landscape and improve community health and vitality. For several years, CleanWays has offered "steps and bridges" cleanups that target areas around the city steps and bridges known as hotspots for trash and illegal dumping. To learn more, visit www.alleghenycleanways.org.

NORTH LANG STREET (EAST END: HOMEWOOD NORTH)

North Lang Street runs through three separate East End neighborhoods, and its northern terminus ends at a flight of city steps. Climbing this flight of stairs, located at the intersection of Chaucer Street, gradually reveals the historic National Opera House. Built in 1894, the Queen Anne–style mansion became a center of Black community life when William "Woogie" Harris, one of Pittsburgh's first Black millionaires, bought the property in

The North Lang Street city steps lead to the historic National Opera House in Homewood North.

1930. The house rose to national significance in 1941 when Mary Cardwell Dawson, a celebrated musician and educator, established the National Negro Opera Company, the first permanent and longest-running African American opera company in the nation. For decades the house was a haven and cultural hub for Black luminaries. The property is currently undergoing rehabilitation to return it to its former glory.

Where to find it: North Lang and Chaucer Streets.

GAZZAM STREET
(EAST END: WEST OAKLAND/TERRACE VILLAGE)

The hillsides of West Oakland and Terrace Village in the East End may be some of the most transformed locations in the city when compared to historical maps of the early twentieth century. Once a densely inhabited area with streets, alleys and city steps weaving throughout the hillsides, today's development is primarily limited to areas above and below, leaving the hillside to return to its natural state. While most of the built environment is gone, the hillside contains the remains of many concrete city steps. Exploring this area, especially in winter, is quite an adventure. The best access point is through Gazzam Street, off Kirkpatrick Street, near the Birmingham Bridge. Look for a flight of stairs to the left of the "No Outlet" sign and wander up and along the hillside to Bentley Drive. Hiking along the hillside parallel to Bentley will lead to the Allequippa Street stairs, another long and meandering flight. For those wanting to explore this area thoroughly, use the Pittsburgh Mapping and Historical Site Viewer (see "Resources") to see maps of the area from 1923. Watch out for open drainage holes and respect fences and private property signs. With a bit of tenacity, it is possible to "follow the old roads" over to Robinson Street near Carlow University.

Where to find it: Kirkpatrick and Gazzam Streets.

COLWELL STREET
(EAST END: HILL DISTRICT/CRAWFORD ROBERTS)

The stone slab city steps that connect Colwell Street to Diaz Way and Lombard Street in the Crawford Roberts section of the Hill District are visually striking in their construction and in fantastic condition—the only signs of wear and tear, beyond the rusted railings, are the grooves worn into the stone from 125 years of pedestrian travel. First appearing on historical maps from 1903, these stairs overlooked the Prohibition-era bootlegging operation of Joe Tito, a prominent businessman and owner of Latrobe Brewing Company. The former Tito garage, located at 1818 Colwell, became the Latrobe Brewing Company's first Pittsburgh beer distributorship and was the first-known place where Rolling Rock beer was sold, beginning in 1935. Tito's house, a grand Victorian that preservation advocates spent years lobbying to save, is located around the corner on Fifth Avenue.

Where to find it: Eastern end of Colwell and Dinwiddie Streets.

Chauncey Street
(East End: Hill District/Middle Hill)

The Chauncey Street city steps once bordered Central Amusement Park, an early 1920s ballfield in the Middle Hill and the first professional baseball field in the country to be owned, designed and built by African Americans. Before the Pittsburgh Crawfords and Homestead Grays, two legendary Negro League teams, there were the Pittsburgh Keystones, and Central Park was built for them. The park was commissioned in 1920 by the Keystones' owner, Alexander M. Williams, and designed by the prominent African American architect Louis Arnett Stuart Bellinger. Bellinger, one of only sixty African American architects at the time, would later design what is now known as the New Granada Theatre at 2007 Centre Avenue. While the Keystones didn't last long, the site remained active through 1925, when it was demolished. Surprisingly, given the population density of the Hill District at this time, the lot remains empty to this day.

The Chauncey Street stairs are best accessed from Centre Avenue, with ample parking and public transportation. The ballpark was bounded by Humber Way (north/northwest, third base), Junilla Street (northeast/east, left field), Hallett Street (southeast/south, right field) and Chauncey Street (southwest/west, first base).

Where to find it: Centre Avenue and Chauncey Street.

Juno Street
(East End: Central Oakland)

Given the extreme depopulation of this region in recent decades, it is tempting to imagine every street bursting at the seams back in the day. While

Traveling the wooden Juno Street city steps isn't for the faint-hearted, but it's worth it for the underside view of the Charles Anderson Bridge.

that was true in many areas, not all neighborhoods were overcrowded. A glance at city maps from one hundred years ago shows that life along Juno Street, located deep in the belly of Central Oakland's Panther Hollow, was quiet, with only a few buildings. These days, Juno is no longer a city road, and the few remaining structures are gradually fading into the natural landscape. The wooden flight leading from the parklet at the intersection of Boulevard of the Allies and Parkview Avenue is in disrepair, but that doesn't stop the spot from attracting attention. A careful walk around shows that it's routinely visited by those seeking an insider view of the Anderson Bridge—a massive structure you can only fully appreciate from the viewpoint of this quiet hollow.

Where to find it: East end of the parklet at Boulevard of the Allies and Parkview Avenue.

FRANK CURTO PARK (EAST END: STRIP DISTRICT)

It's hard to imagine a time when the railroad didn't run through Pittsburgh's Strip District. The Pennsylvania Railroad, chartered in 1846 by the Pennsylvania legislature to build a line between Harrisburg and Pittsburgh, developed its first passenger line in 1848, which ran between Philadelphia and Pittsburgh. By 1882, the Pennsylvania Railroad had become the largest railroad, the largest transportation enterprise and the largest corporation in the world. While the railroad tracks created a border between the Strip and the adjacent hillside neighborhoods of Polish Hill, Bedford Dwellings and Crawford Roberts, there were crossing points. These included an incline, also called funiculars (see "History of the Steps: Incline Ups and Downs"), connecting Cliff Street in Crawford Roberts to 17th Street and public stairways leading to the 28th Street bridge in Polish Hill. While the three remaining stairways in this area are challenging to access due to their proximity to the outbound lane of Bigelow Boulevard (30th Street, Morgan Street and Kirkpatrick Street), the flight at Frank Curto Park can be viewed and explored with care. Remarkably, nearly one hundred years ago, several streets lined the hillside between Bigelow Boulevard and the railroad line, extending the few that remain near the 28th Street bridge. Trekking along this hillside area is best done before warm weather. Once the greenery fills in, the hillside becomes dense with vegetation.

Where to find it: Frank Curto Park (car access via the inbound lane of Bigelow Boulevard).

56ᵀᴴ AND 57ᵀᴴ STREETS
(EAST END: UPPER LAWRENCEVILLE/STANTON HEIGHTS)

While the 57th Street city steps is one of the longest flights in the city (345 steps), the top-most segment is closed due to structural decay. However, this East End flight is worth investigating, as it's one of the few remaining locations for orphan houses. While in this area, it's worth taking a side trip down Duncan Street to see the two distinct versions of 56th Street. One is short and quickly reaches a dead end. The other contains three flights of stairs: one is closed and leads down the hillside to Carnegie Street, another is a flight of sidewalk steps and a third leads to Celadine Street.

This little neighborhood first appeared on city maps in 1890, but it's unclear why 56th Street appears in two separate locations. A walk around the block doesn't take long and offers plenty to look at, especially the old, beautiful brickwork designs of the original row houses.

You'll find the 57th Street city steps off Butler Street and accessible by public transportation. However, due to narrow streets, parking closer to Butler Street is advised.

Where to find it: 57th and Christopher Streets.

WACO WAY
(NORTH SIDE: SPRING HILL–CITY VIEW)

For a truly off-the-beaten-path adventure, make a visit to Waco Way. This hidden flight, accessed at the end of Radner Street, once carried workers to and from a soap factory originally built where the highway currently exists. Once known as East Street Valley, this North Side neighborhood was home to residences, shops, industries and civic and religious institutions. Unfortunately, a significant portion of the neighborhood was demolished for the construction of the Parkway North (I-279), which displaced nearly one thousand families throughout the 1970s. In recent years, landslides have caused significant destruction to existing homes and roads. To take a trip back through history via the city steps, walk to the end of Radner Street and look to the left and right of the fire hydrant to see the stairs. Going downhill leads to East Street, but uphill holds the most intrigue. As you reach the top, a Waco Way street sign awaits, soon revealing the remains of

Valette Street with its precariously perched street railings, old retaining walls and foundations and surprisingly little trash. It's an exhilarating walkabout that reinforces the speed and power of Mother Earth to reclaim a forgotten landscape. Visiting in the winter yields the best views because the area becomes very overgrown during warmer weather.

Where to find it: East and Lareda Streets.

GIRDLEY WAY / STAAB WAY
(NORTH SIDE: SPRING HILL–CITY VIEW)

If exploring Waco Way inspired you to see more of the area once known as the East Street Valley neighborhood, continue north on East Street and turn right on Royal Street. St. Boniface Catholic Church, constructed in 1925 and 1926 and added to the National Register of Historic Places in 1981, is on the left. The Girdley Way stairs, currently closed due to structural decay, are next to the retaining wall on Royal. Girdley Way is one long flight with a perpendicular connecting flight (Staab Way) farther up the hillside. No houses remain on the hillside, but utility poles stand tall. In early spring, it's possible to find daffodils and lilacs, the last vestiges of a once-vibrant neighborhood. For those wanting to fully explore this area and develop a greater understanding of the East Street Valley neighborhood, use the Pittsburgh Mapping and Historical Site Viewer (see "Resources") to see maps of the area from 1923. When hiking through these areas, watch out for and avoid open drainage holes and respect fences and private property signs.

Where to find it: East and Royal Streets.

MARVISTA STREET
(NORTH SIDE: CALIFORNIA KIRKBRIDE)

The far end of Marvista Street on the North Side could easily double as the set of a Hollywood post-apocalyptic thriller. Three flights of stairs line the sides of the street, and another two climb into the hillside. An old "handicap parking" sign looks eerily out of place, the only indication that someone lived here in the recent past. One elevated flight of sidewalk steps, closest to California Avenue, has several occupied homes at the start, but as the flight and street continue, residences give way to empty hillsides and lots.

The hillside flights that lead to Winifred and Success Streets are not closed but are structurally compromised in places, making a hillside scramble more appropriate. The small neighborhood at the top of both flights is very much inhabited and filled with older and newer homes. From here, it's a short distance to Union Dale Cemetery, an enjoyable walk on its own. Created in 1846, Union Dale formed part of the rural cemetery movement that swept the country during the 1830s and '40s as churchyard burial grounds in developing cities grew unpleasantly crowded and increasingly valuable for development. The Union Dale Cemetery website (uniondalecemetery.org) has a list of local notables interred on its ninety-six-acre grounds, including short biographies of each. Reading through the stories offers a glimpse into the region's history and the many people who contributed to its development.

Where to find it: California Avenue and Marvista Street.

VINECLIFFE STREET
(SOUTH HILLS: MOUNT WASHINGTON)

The one-mile stretch of Grandview Avenue in the South Hills neighborhood of Mount Washington may be one of the most popular places frequented by residents and visitors. With wide sidewalks, several strategically placed overlooks with jaw-dropping views of Pittsburgh, historic inclines, churches, working inclines and plenty of amenities, it's a great place to explore on foot and very accessible for a range of abilities.

Mount Washington was once home to the legendary one-thousand-step Indian Trail public stairway (see "History of the Steps" chapter). While that flight was demolished in 1935, it's possible to reimagine the stairs by standing at the intersection of Grandview and Shaler Street and looking down and over the hillside below. The hillside along Grandview Avenue is exceptionally steep and prone to landslides. Stay behind the railings and observe at a safe distance.

On the opposite end of Grandview, near the Monongahela Incline, the road rounds a corner and becomes Wyoming Street. As you travel along this Belgian block street, look to the left for the turquoise blue railings that mark the entrance to the Vinecliffe Street city steps. Vinecliffe first appeared on city maps in 1872, and the current flight of city steps that descends the hillside was constructed in 1952 as part of the city steps improvement plan enacted in Pittsburgh following the end of World War II. Until 2007, Vinecliffe Street still had close to a dozen houses, but by 2008, all were reclaimed by a combination of landslides and a prospective hotel development deal that has yet to materialize. Traveling these stairs, especially in the colder months when the greenery has died back, offers an up-close view of the steep hillside terrain. Mountain bike enthusiasts can also visit YouTube for an up-close view of the city steps and a wild, bumpy ride courtesy of Michael Burner's "Mt Washington (Vinecliffe St) Stairs Down to Pittsburgh" video.

Where to find it: Wyoming and Vinecliffe Streets.

Corinth, Journal and Junius Streets (West End: West End Village / Ridgemont)

This off-road adventure may be the most challenging of all because of its limited accessibility and representation on modern-day street maps. Use the Pittsburgh Mapping and Historical Site Viewer (see "Resources") to see maps of the area from 1923, as they provide helpful context and orientation since infrastructure features such as streets, bridges and stairs have changed.

An article from the *Pittsburgh Sun Telegraph* from November 20, 1958, had this to say about the stairs in this area: "If it's just steps you're looking for, without connecting platforms, it's hard to beat those on Chicken Hill in the West End. They have 580 feet of steps and extend to Woodville Avenue."

Today, Journal Street is only one-tenth of a mile long, and the two flights of city steps are close to the middle on either side of the street. The flight connecting to Junius Street is uphill, with the other flight leading to the train tracks and Corinth Street. Corinth is still listed on current city street maps but is no longer maintained, as all existing houses in this area have been demolished. Another flight of steps constructed later (and not appearing on the 1923 map) follows along Endness Street to Woodville Avenue. It's important to note that residential and commercial properties line Woodville; please respect all fences and private property signs. This area can only be visited in the winter because the greenery is substantial, and an unobstructed view makes the old roads and pathways more visible. You may still find a few streetlights standing, a sign of neighborhoods past.

The remains of the uniquely designed McCartney Street city steps are slowly being reclaimed by nature.

Where to find it: Using public transportation, the 38 bus stops at Greentree Road and Junius Street. If driving, off-road street parking may be available along Junius, Journal or Adolph Streets. These roads are very narrow, so use caution and be mindful of residents' parking places.

If the scramble to the Corinth, Journal and Junius Streets stairs is too extreme for your comfort but you'd still like to explore this area of the West End, visit the McCartney Street city steps that connect to Noblestown Road (also known as Lincoln Highway or Route 60). This flight is also best viewed in winter because greenery overtakes the hillside. At the time of its construction in the mid-twentieth century, it provided easy pedestrian access to West End Park. This flight traverses the hillside through a narrow right-of-way and, because of this constraint, employs four turnouts. The design is unique, and no other flight in the city is constructed in this manner. Today, the hillside is eroded and the flight is structurally unsound, so viewing from the street may be ideal.

Where to find it: The western end of McCartney Street, close to the end of the road.

MARIGOLD LANE
(EAST END: GLEN HAZEL)

For the sharp-eyed, several flights of city steps are present along the overgrown hillsides lining Johnston Avenue in Glen Hazel. These stairs once led to the federally owned 1,001-unit Glen Hazel Heights defense workers' housing project. Before World War II, the land in this area, known as "Sugar Hill," was privately owned and mined for coal. But by early 1942, families had begun moving into the rapidly constructed three-story buildings along the hilltop, and almost immediately, structural problems and water runoff issues began to plague the entire complex.

Despite the less-than-ideal conditions, residents operated one of Pennsylvania's largest food cooperatives, had access to a Carnegie Public Library branch location and developed an active Resident's Council, which successfully organized a strike against Federal Public Housing Authority rent schedules in 1943.

The poor housing quality and ground instability due to severe undermining worsened with time, and in 1953, the property was transferred to Pittsburgh's

Housing Authority. By 1975, only 228 of the original units were occupied, and those residents were given priority for the current Glen Hazel housing, which lies downhill along Johnston Avenue.

The stairs at the southern end of Marigold Lane present a hillside scramble and a bit of bushwacking, but the persistent will connect to an abandoned section of Rivermont Drive. Here, it's possible to see the old streets, curbs and concrete platforms on which many buildings once rested.

Where to find it: The southern end of Marigold Lane or Rivermont Drive.

THE NEXT STEP IS YOURS

Once you start exploring the walks and sites mentioned in this book, don't be surprised if you become a little obsessed with city steps. All that concrete, wood and steel winding through hillsides and steep streets make a lasting impression. When you find yourself excitedly planning your next walk, scouring historical maps for answers to thorny questions or eagerly talking to friends and co-workers about your recent discoveries, congratulations—you're officially hooked.

City steps enthusiasts are an ever-growing group whose hearts, minds, bodies and souls hold a special place for the stairs. This diverse and inclusive group contains every possible profession, category and classification. The city steps are for all people, and thanks to modern technology like Google

Street View, anyone, regardless of their mobility or location, can experience the joys of their exploration.

The stairs, both old and new, are designed to transport people. But to preserve their relevance in the modern world, they need you. We hope that this book inspires you to travel through the city's neighborhoods, enjoy the views, ponder the sights and search for answers to questions that arise on your journey. The experience can be inspiring when you let it be your guide.

We look forward to crossing paths in the future and learning about your adventures.

—LAURA, CHARLES AND MATT

RESOURCES

S everal online resources offer more information about the city steps, their neighborhoods and their connection to the greater transportation network within the city.

Introducing the Pittsburgh City Archives

Since 2018, employees at the Pittsburgh City Archives have been identifying and cataloguing historical records and essential documents as part of ongoing efforts to improve the management of city records. Before then, the city had no official records management division. Other than its legislation, which the city clerk's office has meticulously archived and stored in a fireproof vault, Pittsburgh had no one keeping track of its documents.

Most cities, especially those of Pittsburgh's size, established systems of preserving and organizing municipal records long ago. Cleveland's archives were established in 1814, the National Archives in 1934 and Philadelphia's in 1952.

For much of Pittsburgh's history, each city office independently managed its information without professional guidance, policies or resources for preservation and public access. Every year, the volume of information expanded, while storage mechanisms became increasingly haphazard. Unfortunately, this means irreplaceable documents from previous mayoral administrations and obsolete departments have been misplaced, discarded or stolen. Until recently, the public and historians have been unable to review surviving records that detail city operations.

In 2012, former city councilor Patrick Dowd started advocating for preservation efforts when he discovered records of the city's history in a damp basement. Some argued that the ongoing lack of records management control exposed the city to serious risks associated with efficiency, legal compliance and litigation discovery. Addressing these concerns led to the creation of the Records Management Division within the city clerk's office. Along with other duties, the division manages the city archives, making administrative and historical records accessible to the public and city employees for research.

Current holdings in the city archive include records from city council, public works, city planning, correspondence, meeting minutes, reports, photographs, maps and much more. Highlights include the Urban Redevelopment Authority (URA) Library Collection and digitized Pittsburgh Municipal Records from 1868 to 2000, which include city council meeting minutes, proposed legislation, correspondence and more. The municipal record collection was instrumental in learning how the city built, repaired and funded the city steps throughout the late nineteenth and early twentieth centuries.

This growing treasure-trove of previously inaccessible records is updated when collections are processed and ready for public access. The current catalogue is available by visiting https://pittsburghcityarchives. libraryhost.com.

MAPS

BikePGH: Pittsburgh Bike Map
bikepgh.org
The Pittsburgh Bike map depicts the different types of cycling and pedestrian infrastructure available in Pittsburgh and offers suggestions for safe routes between some of the city's most highly traveled neighborhoods.

City of Pittsburgh Neighborhoods Map
gis.pittsburghpa.gov/pghneighborhoods
Great to use in conjunction with Google Maps for a visual aid in locating Pittsburgh's ninety neighborhoods.

GOOGLE MAPS STREET VIEW
www.google.com/maps
While Google Maps shows most flights of city steps, it's not always easy to assess their current condition. Using the Street View feature found in the bottom right-hand corner, drag and drop the human icon to a street for a 360-degree view. You'll find historical views by looking to the top left corner and selecting the down arrow next to the Street View month and year. Pittsburgh's Street View dates to 2007, so this feature shows what an area looks like in different seasons and also changes that have occurred over time.

G.M. HOPKINS COMPANY MAPS
historicpittsburgh.org/maps-hopkins
Offered through the University of Pittsburgh's library system, the Hopkins maps show Pittsburgh as it was in the years 1872–1940.

PITTSBURGH MAPPING AND HISTORICAL SITE VIEWER
arcgis.com/apps/View/index.html?appid=63f24d1466f24695bf9dfc5
   bf6828126
If you've ever wondered how a neighborhood or street has changed over the years, this interactive map can help. Select a map from a past year and layer it with a more current street map. The results are fascinating. (As this URL is cumbersome, you can also search on "Pittsburgh Historic Maps" to arrive at this location.)

StepsPGH
pittsburghpa.gov/citysteps
This website is part of the City of Pittsburgh's Department of Mobility and Infrastructure and shows the location of almost all the stairways that exist today. It's a great resource.

WESTERN PENNSYLVANIA REGIONAL DATA CENTER
data.wprdc.org/dataset/city-steps
The official City of Pittsburgh database is where you can access or download the city's comprehensive list of all the public stairways in the city. This list is routinely updated.

ADDITIONAL NEIGHBORHOOD WALKING TOURS

FINEVIEW FITNESS TRAIL
mapmyrun.com/routes/view/298817179
It's called Fineview for a reason. When you need to catch your breath from the stairs and elevation, you can look out over downtown. For extra credit, jump off the official trail at Rising Main to test your superhuman powers.

To Heaven and Back: Troy Hill
youtu.be/lx-nhRgVNMY
Created by Erin Anderson and Danny Bracken in collaboration with Troy Hill Citizens Council, this 2019 video walk showcases the many city steps in Troy Hill. A map of the walk is available at https://troyhillsteps.com.

Southside Slopes Historic Route Series
southsideslopes.org/steptrek/historic-steptrek-routes
If you're a fan of the annual Step Trek, check out these South Side Slopes walking routes from past events.

Stepping Out in Pittsburgh's Neighborhoods
mis-steps.com/resources/city-steps-walking-tours
Pittsburgher Stuart Putnam became fascinated with the city steps and wrote a series of short walking tours exploring some of his favorite neighborhoods.

Books

Allegheny City Society. *Allegheny City, 1840–1907.* Charleston, SC: Arcadia Publishing, 2007.

Ansberry, Clare. *The Women of Troy Hill: The Back-Fence Virtues of Faith and Friendship.* New York: Harcourt, 2000.

ASCE Pittsburgh. *Engineering Pittsburgh: A History of Roads, Rails, Canals, Bridges & More.* Charleston, SC: The History Press, 2008.

Aurand, Martin. *The Spectator and the Topographical City.* Pittsburgh, PA: University of Pittsburgh Press, 2014.

Beck, Nancy J. Kimmerle. *Mount Washington and Duquesne Heights.* Charleston, SC: Arcadia Publishing, 2007.

Boehmig, Stuart P. *Pittsburgh's South Side.* Charleston, SC: Arcadia Publishing, 2006.

Doherty, Donald. *Pittsburgh's Inclines.* Charleston, SC: Arcadia Publishing, 2018.

Donoughe, Ron. *Pittsburgh: 90 Neighborhoods.* Pittsburgh, PA: Broudy Printing, 2015.

———. *The Ways of Pittsburgh.* Pittsburgh, PA, 2023.

Glasco, Laurence A., and Christopher Rawson. *August Wilson: Pittsburgh Places in His Life and Plays.* Pittsburgh, PA: Pittsburgh History & Landmarks Foundation, 2015.

Iacone, Audrey, Anna Loney, Nate Marini and Robert Thomas. *Beechview*. Charleston, SC: Arcadia Publishing, 2005.

King, Maxwell, and Louise Lippincott. *American Workman: The Life and Art of John Kane*. Pittsburgh, PA: University of Pittsburgh Press, 2022.

Kleinberg, S.J. *The Shadows of the Mills: Working Class Families in Pittsburgh, 1870–1907*. Pittsburgh, PA: University of Pittsburgh Press, 1989.

Muller, Edward K., and Joel A. Tarr. *Making Industrial Pittsburgh Modern: Environment, Landscape, Transportation, Energy, and Planning*. Pittsburgh, PA: University of Pittsburgh Press, 2019.

O'Neill, Brian. *The Paris of Appalachia: Pittsburgh in the Twenty-First Century*. Pittsburgh, PA: Carnegie Mellon University Press, 2009.

Rooney, Dan, and Carol Peterson. *Allegheny City: A History of Pittsburgh's North Side*. Pittsburgh, PA: University of Pittsburgh Press, 2014.

Yanosko, James W., and Edward W. Yanosko. *Around Troy Hill, Spring Hill, and Reserve Township*. Charleston, SC: Arcadia Publishing, 2011.

INDEX

ABOUT THE AUTHORS

Charles, Laura and Matt at the top of the Middle Street city steps in East Allegheny. *Photo by Kelly Klabnik.*

MATTHEW JACOB works for the Pittsburgh Water and Sewer Authority, managing its asset management software system. He worked for the City of Pittsburgh from 2014 to 2022, inventorying and mapping thousands of city assets such as steps, retaining walls, trails, buildings, bridges and more. Matthew holds a degree in history from Temple University and a master's degree in public management from Carnegie Mellon University.

CHARLES SUCCOP works in the Pittsburgh City Archives as the City Archivist and is the local historian behind the @pghthenandnow Instagram account. He graduated from Appalachian State University, where he studied community and regional planning and geography, and the University of Pittsburgh, where he received a Master of Library and Information Science degree.

LAURA ZUROWSKI is a technical writer at the University of Pittsburgh. Since 2017, she has published the site and blog Mis.Steps: Our Missed Connections with Pittsburgh's City Steps, a public space documentation project of Pittsburgh's public stairways. Laura graduated from Emerson College, where she studied marketing and communication, and Harvard University's Graduate School of Education, where she received a master's degree in education, planning and social policy.